This '

Organic
Laboratory Techniques

Ralph J. Fessenden
University of Montana

(The Late) Joan S. Fessenden
University of Montana

With Contributions by John A. Landgrebe

Brooks/Cole Publishing Company
Pacific Grove, California

Brooks/Cole Publishing Company
A Division of Wadsworth, Inc.

Printed in the United States of America

10 9 8 7 6 5 4 3 2

Library of Congress Cataloging-in-Publication Data
Fessenden, Ralph J., [date]
    Organic laboratory techniques/ Ralph J. Fessenden, Joan S.
Fessenden; with contributions by John A. Landgrebe. —2nd ed.
        p.   cm.
    Includes index.
    ISBN 0-534-20160-1
    1. Chemistry, Organic—Laboratory manuals.   I. Fessenden, Joan S.
II. Landgrebe, John A.   III. Title.
QD261.F466 1993
547′.0078—dc20   92-38632
                                                                 CIP

Sponsoring Editor: *Maureen A. Allaire*
Editorial Assistant: *Beth Wilbur*
Production Editor: *Nancy L. Shammas*
Manuscript Editor: *Carol Beal*
Permissions Editor: *Karen Wootten*
Interior and Cover Design: *Sharon L. Kinghan*
Cover Photo: *Comstock Inc., Mike & Carol Warner*
Art Coordinator: *Lisa Torri*
Typesetting: *Science Typographers*
Printing and Binding: *Malloy Lithographing, Inc.*

Preface

· ·

Organic Laboratory Techniques, second edition, is a supplemental text for the introductory organic laboratory course in which the experiments are supplied by the instructor or in which the students work independently. In this text, the standard laboratory techniques are described, along with suggestions for experiments.

Since the techniques presented here are used in laboratory work by students of biochemistry, molecular biology, and cell biology, this text could be used in selected life science laboratory courses as well. It is also my hope that *Organic Laboratory Techniques* finds its way into the professional library of every science student.

The general order of presentation is isolation and purification techniques, refractive index and chromatography, setting up and carrying out reactions, infrared and NMR spectroscopy, and use of the chemical literature. The order in which the techniques are used is flexible. For example, a student could use Technique 11 (refractive index) along with distillation (Techniques 5–8).

Each technique includes a brief theoretical discussion. Because many students in this course have not yet had a course in physical chemistry, we have kept these discussions fairly general. The steps used in each technique, along with common problems that might arise, are discussed in detail. Supplemental or related procedures, suggested experiments, and suggested readings are included for many of the techniques. Each chapter ends with a set of study problems that primarily stress the practical aspects of the technique.

We have emphasized safety throughout all the techniques. Safety notes are included with each technique where appropriate. Appendix III ("Toxicology of Organic Compounds") is included to alert the student to the possible toxicities of even common compounds.

This second edition has expanded or added information on the following topics.

Safety and Waste Disposal: The introduction to the text on safety has been expanded to include a discussion of locating chemical safety information. Emphasis has been placed on chemical container labels and on Material Safety Data Sheets (MSDS). Appendix IV has been added to deal with the disposal of chemical wastes from student laboratories.

Microscale: A chapter (Technique 10) has been added for those instructors who wish to include microscale experiments. The basic microscale procedures—material transfer, crystallization, distillation, and extraction—have been included. The discussion of these techniques is not exhaustive. The intent is to provide the instructor with a nucleus of information upon which to build.

Additional chromatographic techniques: Flash chromatography and an expanded discussion of HPLC have been added to the chapter on column chromatography (Technique 13).

Spectroscopy: All theory and spectral interpretation has been removed from the text. This material, for the most part, is redundant, since these topics are covered in lecture texts. In addition, there are several short laboratory texts that specialize in these topics. (See the references given at the end of Technique 16.) The space has been used to present an expanded discussion on sampling techniques for spectroscopy (IR, NMR, and UV/visible). In addition, the chapter now covers some instrumental aspects of NMR spectroscopy: ringing, spinning sidebands, and high– and low–field strength spectrometers. The NMR section also includes a discussion of shift reagents and deuterium.

Miscellaneous changes: The one experiment, a Grignard reaction, found in the first edition has been dropped. The reason for this deletion is, in part, to focus text on laboratory techniques only. Other changes include additional background on acid–base reactions used in chemical extraction (Technique 3) and an increased emphasis on on-line computer searches in Technique 17 ("Introduction to the Chemical Literature"). Last, but not least, I hope the numerous small changes made throughout the text have improved its readability.

I am indebted to Joyce Brockwell, Northwestern University; Patricia Feist, University of Colorado-Boulder; Paul McMaster, College of the Holy Cross; James Pavlik, Worcester Polytechnic Institute; and Jan Simek, California Polytechnic State University who have provided extensive and extraordinarily useful comments. I am particularly indebted to John Landgrebe for his help with the chapter on microscale techniques and for figures and sections scattered throughout the text.

Ralph J. Fessenden

Contents

Introduction to the Organic Laboratory

1 Safety in the Laboratory

Organic chemistry is an experimental science. Our understanding of organic chemistry is mainly the result of laboratory observation and testing. For this reason, the laboratory is an important part of a student's education in organic chemistry.

Because of the nature of organic compounds, the organic chemistry laboratory is generally more dangerous than the inorganic chemistry laboratory. Many organic compounds are volatile and flammable. Some can cause chemical burns; many are toxic. Some can cause lung damage, some can lead to cirrhosis of the liver, and some are *carcinogenic* (cancer causing). Yet organic chemists generally live as long as the rest of the population because they have learned to be careful. When working in an organic laboratory, you must always think in terms of safety.

Summary of Safety Rules

It may happen that you are confronted with a laboratory accident and cannot remember exactly what to do. In such a situation, just remember the following:

In the case of a spill: WASH!
In the case of a fire: GET OUT!

In either case, your instructor or someone in a calmer frame of mind can then decide how to handle the situation.

A. Personal Safety

(1) Using Common Sense

Most laboratory safety precautions are nothing more than common sense. The laboratory is not the place for horseplay. Do not work alone in the laboratory. Do not perform unauthorized experiments. Do not sniff, inhale, touch, or taste organic compounds, and do not pipet them by mouth. Wipe up all spilled chemicals, using copious amounts of water to wash up spilled acids and bases. Neutralize residual spilled acid with sodium bicarbonate and spilled base with diluted acetic acid. Do not put dangerous chemicals in the waste crock—the janitor may become injured. Do not pour chemicals down the sink—the environment will be injured. Instead, use the containers provided for chemical disposal.

When working in the laboratory, wear suitable clothing. Jeans and a shirt with rolled-up sleeves, plus a rubber lab apron or cloth lab coat, are ideal. Do not wear your best clothing—laboratory attire usually acquires many small holes from acid splatters and may also develop a distinctive aroma. Loose sleeves can sweep flasks from the laboratory bench, and they present the added hazard of easily catching on fire. Long hair should be tied back. Broken glass sometimes litters the floor of a laboratory; therefore, always wear shoes. Sandals are inadequate because they do not protect the feet from spills. Wash your hands frequently, and always wash them before leaving the laboratory, even to go to the rest room.

Because of the danger of fires, smoking is prohibited in laboratories. Because of the danger of chemical contamination, food and drink also have no place in the laboratory. On the first day of class, familiarize yourself with the locations of the fire extinguishers, fire blanket, eyewash fountain, and shower.

(2) Safety Glasses

Chemicals splashed in the eyes can lead to blindness; therefore, it is imperative that you wear **safety glasses,** or better yet, **safety goggles.** Wear them *at all times*, even if you are merely adding notes to your laboratory notebook or washing dishes. You could be an innocent victim of your lab partner's mistake, who might inadvertently splash a corrosive chemical in your direction. In the case of particularly hazardous manipulations, you should wear a **full-face shield** (similar to a welder's face shield). Your instructor will tell you when this is necessary.

Contact lenses should not be worn, even under safety glasses. The reason for this rule is that contact lenses cannot always be removed quickly if a chemical gets into the eye. A person administering first aid by washing your eye might not even realize that you are wearing contact lenses. In addition, "soft" contact lenses can absorb harmful vapors. If contact lenses are absolutely necessary, properly fitted goggles must be worn. Also, inform your laboratory instructor and neighbors that you are wearing contact lenses.

B. Laboratory Accidents

(1) Chemicals in the Eyes

If a chemical does get into your eye, flush it with gently flowing water for 15 minutes. Do not try to neutralize an acid or base in the eye. Because of the natural tendency for the eyelids to shut when something is in the eye, *they must be held open during the washing*. If there is no eyewash fountain in the laboratory, a piece of rubber tubing attached to a faucet is a good substitute. Do not take time to put together a fountain if you have something in your eye, however! Either splash your eye (held open) with water from the faucet immediately or lie down on the floor and have someone else pour a gentle stream of water into your eye. *Time* is important. The sooner you can wash a chemical out of your eye, the less the damage will be.

After the eye has been flushed, medical treatment is strongly advised. For any corrosive chemical, such as sodium hydroxide, prompt medical attention is imperative!

(2) Chemical Burns

Any chemical (whether water-soluble or not) spilled onto the skin should be washed off immediately with soap and water. The detergent action of the soap and the mechanical action of washing remove most substances, even insoluble ones. If the chemical is a strong acid or base, rinse the splashed area of the skin with *lots and lots of cool water*. Strong acids on the skin usually cause a painful stinging. Strong bases usually do not cause pain, but they are extremely harmful to tissue. Always wash carefully after using a strong base.

If chemicals are spilled on a large area of the body, wash them off in the safety shower. If the chemicals are corrosive or can be absorbed through the skin, remove contaminated clothing so that the skin can be flushed thoroughly. If chemical burns result, the victim should seek medical attention.

(3) Heat Burns

Minor burns from hot flasks, glass tubing, and the like are not uncommon occurrences in the laboratory. The only treatment needed for a very minor burn is holding it under cold water for 5–10 minutes. A painkilling lotion may then be applied. To prevent minor burns, keep a pair of inexpensive, loose-fitting cotton gloves in your laboratory locker to use when you must handle hot beakers, tubing, or flasks.

A person with a serious burn, as from burned clothing, is likely to go into shock. He or she should be made to lie down on the floor and kept warm with the fire blanket or with a coat. Then, an ambulance should be called. Except to extinguish flames or to remove harmful chemicals, do not wash a serious burn and do not apply any ointments. However, cold compresses on a burned area will help dissipate heat.

(4) Cuts

Minor cuts from broken glassware are another common occurrence in the laboratory. These cuts should be flushed thoroughly with cold water to remove any chemicals or slivers of glass. A pressure bandage can be used to stop any bleeding.

Major cuts and heavy bleeding are a more serious matter. The injured person should lie down and be kept warm in case of shock. A pressure bandage (such as a folded, clean dish towel) should be applied over the wound and the injured area elevated slightly, if possible. An ambulance should be called immediately.

The use of a tourniquet is no longer advised. Experience has shown that cutting off all circulation to a limb may result in gangrene.

(5) Inhalation of Toxic Substances

A person who has inhaled vapors of an irritating or toxic substance should be removed immediately to fresh air. If breathing stops, artificial respiration should be administered and an emergency medical vehicle called.

C. Laboratory Fires

(1) Avoiding Fires

Most fires in the laboratory can be prevented by the use of common sense. Before lighting a match or burner, check the area for flammable solvents. Solvent fumes are heavier than air and can travel along a benchtop or a drainage trough in the bench. These heavy flammable fumes can remain in sinks or wastebaskets for days. While it is indeed true that a flammable solvent should not have been discarded in the sink or wastebasket, it is always possible that some inconsiderate fool has done so. Therefore, do not discard hot matches, even if extinguished, or any other hot substances in sinks or wastebaskets.

Whenever you use a flammable solvent, extinguish all flames in the vicinity beforehand. Always cap solvent bottles when not actually in use. Do not boil away flammable solvents from a mixture except in the fume hood. Place solvent-soaked filter paper in the fume hood to dry before discarding it in a waste container. Spilled solvent should not be allowed simply to evaporate. If a solvent is spilled, clean it up immediately with paper towels, which should be placed in the hood to dry.

Solvents should never be poured into a drainage trough (which is for water only). Because of environmental concerns, solvents should be disposed of only in containers provided for solvent disposal. In general, these disposal containers are located in the fume hood in the laboratory.

(2) Extinguishing Fires

In case of even a small fire, tell your neighbors to leave the area and notify the instructor. A fire confined to a flask or beaker can be smothered with a watch glass or large beaker placed over the flaming vessel. (Try not to drop a flaming flask—this will splatter burning liquid and glass over the area.) All burners in the vicinity of a fire should be extinguished, and all containers of flammable materials should be removed to a safe place in case the fire spreads.

For all but the smallest fire, the laboratory should be cleared of people. It is better to say loudly, "Clear the room," than to scream "Fire!" in a panicky voice. If you *hear* such a shout, do not stand around to see what is happening, but stop whatever you are doing and walk immediately and purposefully toward the nearest clear exit.

Many organic solvents float on water; therefore, water may serve only to spread a chemical fire. Some substances, like sodium metal, explode on contact with water. For these reasons, water should not be used to extinguish a laboratory fire; instead, a *carbon dioxide* or *powder fire extinguisher* should be used.

If a fire extinguisher is needed, it is best to clear the laboratory and allow the instructor to handle the extinguisher. Even so, you should acquaint yourself with the location, classification, and operation of the fire extinguishers on the first day of class. Inspect the fire extinguishers. Find the sealing wire (indicating that the extinguisher is fully charged) and the pin that is used to break this sealing wire when the extinguisher is needed.

Fire extinguishers usually spray their contents with great force. To avoid blowing flaming liquid and broken glass around the room, aim toward the base and to the side of any burning equipment, not directly toward the fire. Once a fire extinguisher has been used, it will need recharging before it is again operable. Therefore, any use of a fire extinguisher must be reported to the instructor.

(3) Extinguishing Burning Clothing

If your clothing catches fire, walk (do not run) to the shower if it is close by. If the shower is not near, lie down, roll to extinguish the flames, and call for help.

A clothing fire may be extinguished by having the person roll in a fire blanket. The rolling motion is important because a fire can still burn under the blanket. Wet towels can also be used to extinguish burning clothing. A burned person should be treated for shock (kept quiet and warm). Medical attention should be sought.

D. Handling Chemicals

(1) Acids and Bases

To prevent acid splatters, *always add concentrated acids to water* (*never add water to acids*). Concentrated sulfuric acid (H_2SO_4) should be added to ice water or crushed ice because of the heat generated by the mixing. Do not pour acids down the drain without first diluting them (by adding them to large amounts of water) and then neutralizing them. Strong bases should also be diluted and neutralized before discarding. If you splash an acid or strong base on your skin, wash with copious amounts of water, as described in the section on chemical burns. Concentrated hydrochloric acid (HCl) and glacial acetic acid (CH_3CO_2H) present the added hazard of extremely irritating vapors. These two acids should be used only in the fume hood.

Sodium hydroxide (NaOH) is caustic and can eat away living tissue. As a solid (usually pellets), it is deliquescent; a pellet that is dropped and ignored will form a dangerous pool of concentrated NaOH. Pick up spilled pellets while wearing plastic gloves or by using a piece of paper, neutralize them, and then flush the neutralized mixture down the drain with large amounts of water.

Aqueous ammonia ("ammonium hydroxide") emits ammonia (NH_3) vapors and thus should be used only in the fume hood.

(2) Solvents

Organic solvents present the double hazard of flammability and toxicity (both short-term and cumulative). (Table III.1, page 240, in Appendix III lists the toxic levels and allowable limits of some common solvents.) *Diethyl ether* ($CH_3CH_2OCH_2CH_3$) and *petroleum ether* (a mixture of alkanes) are both very volatile (have low boiling points) and extremely flammable. These two solvents should never be used in the vicinity of a flame, and they should be boiled only in the hood. *Carbon disulfide* (CS_2), which is now rarely used in the organic laboratory, is uniquely hazardous. Its ignition temperature is under 100°, the boiling point of water; therefore, fires can result even from its contact with a steam pipe. *Benzene* (C_6H_6) is flammable and also extremely toxic. It can be absorbed through the skin, and long-term exposure is thought to cause cancer. Benzene should be used as a solvent only when absolutely necessary (and then handled with great care to avoid inhalation, splashes on the skin, or fire). In most cases *toluene* can be substituted for benzene. Although toluene is flammable, it is less toxic than benzene.

Most halogenated hydrocarbons, such as *carbon tetrachloride* (CCl_4) and *chloroform* ($CHCl_3$), are toxic, and some are carcinogenic. Halogenated hydrocarbons tend to accumulate in the fatty tissues in living systems, instead of being detoxified and excreted, as most poisons are. In repeated small doses, they are associated with chronic poisoning and damage to the liver and kidneys. If either carbon tetrachloride or chloroform must be used, it should be handled in the fume hood.

Because of the dangers inherent with all organic solvents, they should always be handled with respect. Solvent vapors should not be inhaled, and solvents should never be tasted or poured on the skin. Wash any splashes on your skin immediately with soap and water. Keep solvent bottles tightly capped. Always heed precautions to avoid fires.

E. Safety Information

(1) Labeling of Chemical Containers Found in Student Laboratories

There has been a concerted effort in recent years to provide safety information on all chemical containers. This effort has been very successful for those containers that are sold by chemical suppliers but less so for those groups, such as schools, that buy bulk chemicals and repackage them for their own internal use, such as for use in a classroom.

If the chemical container used in your classroom originated with a chemical vendor, it will have a *precautionary label*. A typical precautionary label is shown in Figure 1. These labels provide at a glance a good idea of the hazards associated with the chemical inside the container. Looking at the acetone label in Figure 1, we can quickly see that acetone has a slight health hazard, that it has a significant fire hazard, that its chemical reactivity hazard is moderate, and that its skin contact hazard is slight. Contrast the acetone label with that for nitric acid, also given in Figure 1. You can quickly see that nitric acid is a far more dangerous chemical. You are alerted to the facts that nitric acid is a severe poison and has an extreme contact hazard, but it is not a fire hazard.

In addition to giving safety information, the label also makes recommendations for safety equipment use. Notice that the acetone label recommends the use of safety glasses when working with acetone, but the nitric acid label recommends a full-face shield and goggles.

Below the colored hazard and protection information, in finer print, is a wealth of other safety facts concerning these two compounds. It is well worth your time to become a label reader.

(2) Material Safety Data Sheets (MSDS)

OSHA (Occupational Safety and Health Administration) regulations prescribe that Material Safety Data Sheets (MSDSs) be made available upon request for every chemical used in an industrial or academic operation. The exact format of these MSDSs vary with the different vendors. However,

protection information

hazard information

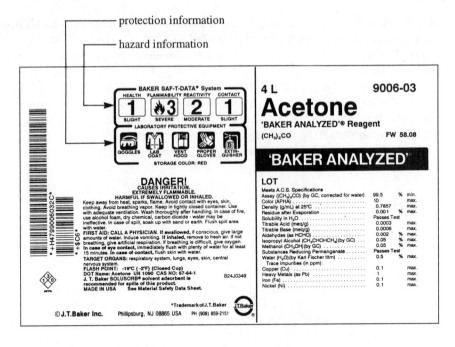

FIGURE 1 Precautionary labels. (Courtesy of J. T. Baker Chemical Inc. Reprinted by permission.)

MSDSs contain an enormous amount of information, and it behooves you to request and study them for every chemical you use.

Two MSDSs from two different manufacturers/distributors follow (Figure 2). In browsing through these documents, you can see the wealth of information that is available to you merely for the asking.

F. Disposal of Chemicals

The techniques for disposing of chemical wastes from classroom laboratories is in a state of change. Appendix IV reviews the current procedures. However, these procedures do vary from laboratory to laboratory. Before you dispose of any chemical, be sure to check with your instructor or the storeroom personnel that you are using the correct procedure.

· · · · · · · · · · ·
2 The Laboratory Notebook

A laboratory notebook serves several purposes. The first is for your own reference. You may think that you will remember everything you do and see in the laboratory; however, many details, such as melting points, boiling points, and weights, are easily forgotten. It is far better to record these details in a well-organized notebook than to try to memorize them. An approved notebook should be used, not little slips of paper, which seem to flutter away when your back is turned.

Another purpose of a laboratory notebook is to allow someone else to review your work or repeat it *exactly*. This facet of experimental work is necessary in research, and it is also necessary in a student laboratory. If a particular experiment does not work well for you, the instructor would want to know why. Your detailed written procedure and observations can give clues to experimental failures.

A research laboratory notebook is also important to help establish the validity of patent claims. Each page or experiment in the notebook must be numbered and dated. If the project is especially promising, the researcher will have his or her notebook pages signed by witnesses. This procedure is usually unnecessary in a student laboratory.

A. The Correct Notebook

The correct notebook for the laboratory is a hardcover, bound book containing lined pages. These are available at bookstores and stationery stores. A loose-leaf or spiral notebook is not satisfactory because pages are easily removed and lost. A separate notebook should be used for each laboratory course.

J.T. Baker Inc.
222 Red School Lane
Phillipsburg, NJ 08865

24-Hour Emergency Telephone 908-859-2151
National Response Center 800-424-8802
Chemtrec 800-424-9300

| National Response In Canada CANUTEC 613-996-6666 |
| Outside U.S. and Canada Chemtrec 202-483-7616 |

MATERIAL SAFETY DATA SHEET

FICHE
SIGNALETIQUE

HOJAS DE DATOS
DE SEGURIDAD

NOTE: CHEMTREC, CANUTEC and National Response Center emergency numbers are to be used only in the event of chemical emergencies involving a spill, leak, fire, exposure or accident involving chemicals. All non-emergency questions should be directed to Customer Service (1-800-JTBAKER) for assistance.

```
S6986 -03              Styrene (Stabilized)              Page: 1
Effective:  05/01/89                              Issued: 09/23/92


J.T.BAKER INC., 222 Red School Lane, Phillipsburg, NJ 08865

==============================================================
                SECTION I - PRODUCT IDENTIFICATION
==============================================================

Product Name:    Styrene (Stabilized)
Common Synonyms: Styrene Monomer; Vinylbenzene; Phenylethylene; Styrol;
                 Cinnamene
Chemical Family: Aromatic Hydrocarbons
```
Formula: $C_6H_5CH:CH_2$
```
Formula Wt.:     104.15
CAS No.:         100-42-5
NIOSH/RTECS No.: WL3675000
Product Use:     Laboratory Reagent
Product Codes:   V091


==============================================================
                  PRECAUTIONARY LABELING
==============================================================
```
BAKER SAF-T-DATA* System

	HEALTH	FLAMMABILITY	REACTIVITY	CONTACT
	2	**3**	**2**	**3**
	MODERATE	SEVERE	MODERATE	SEVERE

Laboratory Protective Equipment

GOGGLES & SHIELD   LAB COAT & APRON   VENT HOOD   PROPER GLOVES   EXTIN-GUISHER

U.S. Precautionary Labeling

DANGER!
FLAMMABLE. CAUSES BURNS. HARMFUL IF SWALLOWED OR INHALED.
Keep away from heat, sparks, flame. Do not get in eyes, on skin, on clothing.
Avoid breathing vapor. Keep in tightly closed container. Use with adequate
ventilation. Wash thoroughly after handling. In case of fire, use alcohol
foam, dry chemical, carbon dioxide - water may be ineffective. Flush spill
area with water spray.

Continued on Page: 2

FIGURE 2   Material Safety Data Sheets (MSDSs) for styrene and *p*-xylene. (Courtesy of J. T. Baker Chemical Inc. and EM Science. Reprinted by permission.)

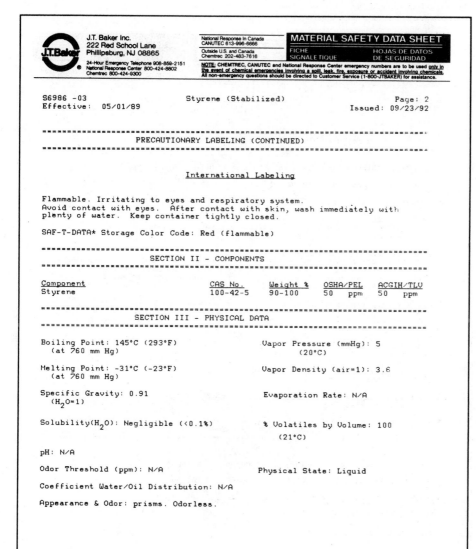

FIGURE 2 *(continued)*

J.T. Baker Inc.
222 Red School Lane
Phillipsburg, NJ 08865

24-Hour Emergency Telephone 908-859-2151
National Response Center 800-424-8802
Chemtrec 800-424-9300

National Response in Canada
CANUTEC 613-996-6666

Outside U.S. and Canada
Chemtrec 202-483-7616

MATERIAL SAFETY DATA SHEET
FICHE
SIGNALETIQUE
HOJAS DE DATOS
DE SEGURIDAD

NOTE: CHEMTREC, CANUTEC and National Response Center emergency numbers are to be used only in the event of chemical emergencies involving a spill, leak, fire, exposure or accident involving chemicals. All non-emergency questions should be directed to Customer Service (1-800-JTBAKER) for assistance.

S6986 -03   Styrene (Stabilized)   Page: 3
Effective: 05/01/89   Issued: 09/23/92

```
======================================================================
              SECTION IV - FIRE AND EXPLOSION HAZARD DATA
======================================================================
```

Flash Point (Closed Cup): 31°C (88°F)   NFPA 704M Rating: 2-3-2

Autoignition Temperature: 489°C (914°F)

Flammable Limits: Upper - 6.1 %   Lower - 1.1 %

Fire Extinguishing Media
    Use alcohol foam, dry chemical or carbon dioxide. (Water may be
    ineffective.)

Special Fire-Fighting Procedures
    Firefighters should wear proper protective equipment and self-contained
    breathing apparatus with full facepiece operated in positive pressure
    mode. Move containers from fire area if it can be done without risk. Use
    water to keep fire-exposed containers cool.

Unusual Fire & Explosion Hazards
    Vapors may flow along surfaces to distant ignition sources and flash back.
    Closed containers exposed to heat may explode. Contact with strong
    oxidizers may cause fire.

Toxic Gases Produced
    carbon monoxide, carbon dioxide

Explosion Data-Sensitivity to Mechanical Impact
    None identified.

Explosion Data-Sensitivity to Static Discharge
    None identified.

```
======================================================================
                    SECTION V - HEALTH HAZARD DATA
======================================================================
```

Threshold Limit Value (TLV/TWA):   215 mg/m³   (50 ppm)

Short-Term Exposure Limit (STEL): 425 mg/m³   (100 ppm)

Permissible Exposure Limit (PEL):   (100   ppm)

PEL (Ceiling) = 200 ppm.

Toxicity of components

Continued on Page: 4

FIGURE 2   (continued)

```
┌─────────────────────────────────────────────────────────────────────────────┐
```

J.T. Baker Inc.
222 Red School Lane
Phillipsburg, NJ 08865

24-Hour Emergency Telephone 908-859-2151
National Response Center 800-424-8802
Chemtrec 800-424-9300

National Response in Canada
CANUTEC 613-996-6666

Outside U.S. and Canada
Chemtrec 202-483-7616

MATERIAL SAFETY DATA SHEET

FICHE
SIGNALE TIQUE

HOJAS DE DATOS
DE SEGURIDAD

NOTE: CHEMTREC, CANUTEC and National Response Center emergency numbers are to be used only in the event of chemical emergencies involving a spill, leak, fire, exposure or accident involving chemicals. All non-emergency questions should be directed to Customer Service (1-800-JTBAKER) for assistance.

S6986 -03   Styrene (Stabilized)   Page: 4
Effective: 05/01/89   Issued: 09/23/92

===
                   SECTION V - HEALTH HAZARD DATA (CONTINUED)
===

Oral Rat LD_{50} for Styrene   5000 mg/kg
Oral Mouse LD_{50} for Styrene   316 mg/kg
Intraperitoneal Mouse LD_{50} for Styrene   660 mg/kg
Intravenous Mouse LD_{50} for Styrene   90 mg/kg
Carcinogenicity: NTP: No   IARC: No   Z List: No   OSHA Reg: No

Carcinogenicity
     None identified.

Reproductive Effects
     None identified.

Effects of Overexposure

     INHALATION: excessive inhalation respiratory system, irritation, may
                       cause pulmonary edema, narcosis

     SKIN CONTACT: burns

     EYE CONTACT: irritation, burns

     SKIN ABSORPTION: none identified

     INGESTION: irritation and burns to mouth and stomach

     CHRONIC EFFECTS: central nervous system depression

Target Organs
     central nervous system, respiratory system, eyes, skin

Medical Conditions Generally Aggravated by Exposure
     none identified

Primary Routes of Entry
     inhalation, ingestion, eye contact, skin contact

Emergency and First Aid Procedures

     INGESTION: CALL A PHYSICIAN. If swallowed, if conscious, give large
                   amounts of water. Induce vomiting.

     INHALATION: If inhaled, remove to fresh air. If not breathing, give
                   artificial respiration. If breathing is difficult, give
                   oxygen.

                         Continued on Page: 5

FIGURE 2   *(continued)*

J.T. Baker Inc. 222 Red School Lane Phillipsburg, NJ 08865 24-Hour Emergency Telephone 908-859-2151 National Response Center 800-424-8802 Chemtrec 800-424-9300	National Response In Canada CANUTEC 613-996-6666 Outside U.S. and Canada Chemtrec 202-483-7616	**MATERIAL SAFETY DATA SHEET** FICHE HOJAS DE DATOS SIGNALETIQUE DE SEGURIDAD

NOTE: CHEMTREC, CANUTEC and National Response Center emergency numbers are to be used only in the event of chemical emergencies involving a spill, leak, fire, exposure or accident involving chemicals. All non-emergency questions should be directed to Customer Service (1-800-JTBAKER) for assistance.

S6986 -03   Styrene (Stabilized)   Page: 6
Effective: 05/01/89   Issued: 09/23/92

==
            SECTION VII - SPILL & DISPOSAL PROCEDURES (CONTINUED)
==

Disposal Procedure
    Dispose in accordance with all applicable federal, state, and local
    environmental regulations.

EPA Hazardous Waste Number: D001 (Ignitable Waste)·

==
              SECTION VIII - INDUSTRIAL PROTECTIVE EQUIPMENT
==

Ventilation: Use general or local exhaust ventilation to meet TLV
                        requirements.

Respiratory Protection: Respiratory protection required if airborne
                        concentration exceeds TLV. At concentrations up to
                        400 ppm, a chemical cartridge respirator with organic
                        vapor cartridge is recommended. Above this level, a
                        self-contained breathing apparatus is recommended.

Eye/Skin Protection: Safety goggles and face shield, uniform, protective
                        suit, polyvinyl acetate gloves are recommended.

==
              SECTION IX - STORAGE AND HANDLING PRECAUTIONS
==

SAF-T-DATA* Storage Color Code: Red (flammable)

Storage Requirements
    Keep container tightly closed. Store in a cool, dry, well-ventilated,
    flammable liquid storage area. Store below 70°C. Store in light-resistant
    containers.

Special Precautions
    Bond and ground containers when transferring liquid.

==
            SECTION X - TRANSPORTATION DATA AND ADDITIONAL INFORMATION
==

Continued on Page: 7

FIGURE 2 *(continued)*

J.T. Baker Inc. 222 Red School Lane Phillipsburg, NJ 08865	National Response In Canada CANUTEC 613-996-6666	**MATERIAL SAFETY DATA SHEET**
24-Hour Emergency Telephone 908-859-2151 National Response Center 800-424-8802 Chemtrec 800-424-9300	Outside U.S. and Canada Chemtrec 202-483-7616	FICHE HOJAS DE DATOS SIGNALETIQUE DE SEGURIDAD

NOTE: CHEMTREC, CANUTEC and National Response Center emergency numbers are to be used only in the event of chemical emergencies involving a spill, leak, fire, exposure or accident involving chemicals. All non-emergency questions should be directed to Customer Service (1-800-JTBAKER) for assistance.

```
S6986 -03                    Styrene (Stabilized)                  Page: 5
Effective:  05/01/89                                     Issued:  09/23/92
```

==
SECTION V - HEALTH HAZARD DATA (CONTINUED)
==

   SKIN CONTACT: In case of contact, immediately flush skin with plenty of
                 water for at least 15 minutes while removing contaminated
                 clothing and shoes. Wash clothing before re-use.

   EYE CONTACT: In case of eye contact, immediately flush with plenty of
                 water for at least 15 minutes.

SARA/TITLE III HAZARD CATEGORIES and LISTS

Acute: Yes Chronic: Yes Flammability: Yes Pressure: No Reactivity: No

```
Extremely Hazardous Substance:  No
CERCLA Hazardous Substance:     Yes   Contains Styrene (RQ = 1000 LBS)
SARA 313 Toxic Chemicals:       Yes   Contains Styrene
     Generic Class:                   Generic Class Removed from CFR:  7/1/91
TSCA Inventory:                 Yes
```

==
SECTION VI - REACTIVITY DATA
==

Stability: Stable   Hazardous Polymerization: Will not occur

Conditions to Avoid: light, heat, flame, other sources of ignition, air

Incompatibles: strong oxidizing agents, copper, strong acids,
                        metallic salts, polymerization catalysts &
                        accelerators

Decomposition Products: carbon monoxide, carbon dioxide

==
SECTION VII - SPILL & DISPOSAL PROCEDURES
==

Steps to be Taken in the Event of a Spill or Discharge
    Wear self-contained breathing apparatus and full protective clothing. Shut
    off ignition sources; no flares, smoking or flames in area. Stop leak if
    you can do so without risk. Use water spray to reduce vapors. Take up
    with sand or other non-combustible absorbent material and place into
    container for later disposal. Flush area with water.

J. T. Baker SOLUSORB[R] solvent adsorbent is recommended for spills of this
product.

Continued on Page: 6

FIGURE 2   (continued)

J.T. Baker Inc. 222 Red School Lane Phillipsburg, NJ 08865	National Response In Canada CANUTEC 613-996-6666	**MATERIAL SAFETY DATA SHEET**	
	Outside U.S. and Canada Chemtrec 202-483-7616	FICHE SIGNALETIQUE	HOJAS DE DATOS DE SEGURIDAD
24-Hour Emergency Telephone 908-859-2151 National Response Center 800-424-8802 Chemtrec 800-424-9300	**NOTE:** CHEMTREC, CANUTEC and National Response Center emergency numbers are to be used <u>only in the event of chemical emergencies involving a spill, leak, fire, exposure or accident involving chemicals.</u> All non-emergency questions should be directed to Customer Service (1-800-JTBAKER) for assistance.		

```
S6986 -03                      Styrene (Stabilized)                    Page: 7
Effective:  05/01/89                                            Issued: 09/23/92
```

```
-----------------------------------------------------------------------
      SECTION X - TRANSPORTATION DATA AND ADDITIONAL INFORMATION (CONTINUED)
-----------------------------------------------------------------------
```

Domestic (D.O.T.)

```
Proper Shipping Name:  Styrene monomer, inhibited
Hazard Class:          Flammable liquid
UN/NA: UN2055    Reportable Quantity: 1000  LBS.
Labels: FLAMMABLE LIQUID
Regulatory References: 49CFR 172.101; 173.119
```

International (I.M.O.)

```
Proper Shipping Name:  Styrene monomer, inhibited
Hazard Class:          3.3                        I.M.O. Page: 3155
UN: UN2055       Marine Pollutants: No            Packaging Group: II
Labels: FLAMMABLE LIQUID
Regulatory References: 49CFR 172.102; Part 176; IMO
```

AIR (I.C.A.O.)

```
Proper Shipping Name:  Styrene monomer, inhibited
Hazard Class:          3.3
UN: UN2055                                         Packaging Group: II
Labels: FLAMMABLE LIQUID
Regulatory References: 49CFR 172.101; 173.6; Part 175; ICAO/IATA=== We believe
                       the transportation data and references contained herein
                       to be factual and the opinion of qualified experts. The
                       data is meant as a guide to the overall classification
                       of the product and is not package size specific, nor
                       should it be taken as a warranty or representation for
                       which the company assumes legal responsibility.=== The
                       information is offered solely for your consideration,
                       investigation, and verification. Any use of the
                       information must be determined by the user to be in
                       accordance with applicable Federal, State, and Local
                       laws and regulations. See shipper requirements 49CFR
                       172.3 and employee training 49CFR 173.1.

U.S. Customs Harmonization Number: 29025000004
```

```
-----------------------------------------------------------------------
```

```
                        Continued on Page:  8
```

FIGURE 2   *(continued)*

J.T. Baker Inc.
222 Red School Lane
Phillipsburg, NJ 08865

24-Hour Emergency Telephone 908-859-2151
National Response Center 800-424-8802
Chemtrec 800-424-9300

National Response In Canada
CANUTEC 613-996-6666

Outside U.S. and Canada
Chemtrec 202-483-7616

MATERIAL SAFETY DATA SHEET

FICHE
SIGNALETIQUE

HOJAS DE DATOS
DE SEGURIDAD

NOTE: CHEMTREC, CANUTEC and National Response Center emergency numbers are to be used only in the event of chemical emergencies invorving a spill, leak, fire, exposure or accident involving chemicals. All non-emergency questions should be directed to Customer Service (1-800-JTBAKER) for assistance.

S6986 -03   Styrene (Stabilized)   Page: 8
Effective: 05/01/89   Issued: 09/23/92

==

N/A = Not Applicable or Not Available
N/E = Not Established

The information in this Material Safety Data Sheet meets the
requirements of the United States OCCUPATIONAL SAFETY AND HEALTH ACT and
regulations promulgated thereunder (29 CFR 1910.1200 et. seq.) and the
Canadian WORKPLACE HAZARDOUS MATERIALS INFORMATION SYSTEM. This document
is intended only as a guide to the appropriate precautionary handling of
the material by a person trained in, or supervised by a person trained
in, chemical handling. The user is responsible for determining the
precautions and dangers of this chemical for his or her particular
application. Depending on usage, protective clothing including eye and
face guards and respirators must be used to avoid contact with material
or breathing chemical vapors/fumes.
Exposure to this product may have serious adverse health effects. This
chemical may interact with other substances. Since the potential uses
are so varied, Baker cannot warn of all of the potential dangers of use
or interaction with other chemicals or materials. Baker warrants that
the chemical meets the specifications set forth on the label.
BAKER DISCLAIMS ANY OTHER WARRANTIES, EXPRESSED OR IMPLIED WITH REGARD
TO THE PRODUCT SUPPLIED HEREUNDER, ITS MERCHANTABILITY OR ITS FITNESS
FOR A PARTICULAR PURPOSE.
The user should recognize that this product can cause severe injury and
even death, especially if improperly handled or the known dangers of use
are not heeded. READ ALL PRECAUTIONARY INFORMATION. As new documented
general safety information becomes available, Baker will periodically
revise this Material Safety Data Sheet.
Note: CHEMTREC, CANUTEC, and NATIONAL RESPONSE CENTER emergency telephone
numbers are to be used ONLY in the event of CHEMICAL EMERGENCIES involving
a spill, leak, fire, exposure, or accident involving chemicals. All
non-emergency questions should be directed to Customer Service
(1-800-JTBAKER) for assistance.

COPYRIGHT 1992 J.T.BAKER INC.
* TRADEMARKS OF J.T.BAKER INC.
===
Approved by Quality Assurance Department.

                          -- LAST PAGE --

FIGURE 2   (continued)

MATERIAL SAFETY DATA SHEET

EM SCIENCE

Ship To:102924

    ORGANIC LAB TECHNIQUES

1. CHEMICAL PRODUCT AND COMPANY IDENTIFICATION

Manufacturer.............:   Preparation Date.: 03/01/91
                               Date MSDS Printed.: Aug 11, 1992

    EM SCIENCE
    A Division of EM Industries   Information Phone Number.: 609-354-9200
    P.O. Box 70   Hours: Mon. to Fri. 8:30-5
    480 Democrat Rd.   Chemtrec Emergency Number: 800-424-9300
    Gibbstown, N.J. 08027   Hours: 24 hrs a day

Catalog Number(s):
   XX0045

Chemical Name....: p-Xylene
Trade Name.......: p-Xylol; 1,4-dimethylbenzene
Chemical Family..: Aromatic Hydrocarbon
Formula..........: $C_6H_4(CH_3)_2$

Molecular Weight.: 106.17

2. COMPOSITION / INFORMATION ON INGREDIENTS

Component	CAS #	Appr %
p-Xylene	106-42-3	100%

3. HAZARDS IDENTIFICATION

EMERGENCY OVERVIEW
   FLAMMABLE LIQUID AND VAPOR.
   HARMFUL OR FATAL IF SWALLOWED.
   VAPOR HARMFUL.
   IRRITATING TO SKIN, EYES AND MUCOUS MEMBRANES.
   May Cause Damage To Lungs, Liver, Kidneys and Central Nervous System.

Appearance..............:
   colorless liquid; aromatic odor

MSDS #XX0045   Page # 1 (continued on next page)

FIGURE 2   *(continued)*

EM®

POTENTIAL HEALTH EFFECTS (ACUTE AND CHRONIC)

Symptoms of Exposure:
 -Harmful if swallowed
 Irritating on contact with skin, eyes or mucous membranes.
 Vapor irritating to eyes and respiratory passages
 May cause damage to central nervous system, liver, lungs,
 kidneys, blood.

Medical Cond. Aggravated by Exposure:
 Nervous System and Rerpiratory System
 conditions, liver, blood and kidney conditions

Routes of Entry.....................:
 Inhalation, ingestion
Carcinogenicity.....................:
 The material is not listed as a cancer causing agent.

 4. FIRST AID MEASURES

Emergency First Aid:
 GET MEDICAL ASSISTANCE FOR ALL CASES OF OVEREXPOSURE.
 Skin: Wash thoroughly with soap and water.
 Eyes: Immediately flush thoroughly with water for at least 15
 minutes.
 Inhalation: Remove to fresh air; give artificial respiration if
 breathing has stopped.
 Ingestion: Call a physician immediately. ONLY induce vomiting at the
 instructions of a physician. Never give anything by mouth to an
 unconscious person.
 Remove contaminated clothing and wash before reuse.

 5. FIRE FIGHTING MEASURES

Flash Point (F)..........: 81F (cc)
Flammable Limits LEL (%).: 1.10
Flammable Limits UEL (%).: 7.00
Extinguishing Media......:
 Foam, Carbon dioxide, Dry chemical

Fire Fighting Procedures.:
 Wear self-contained breathing apparatus.

Fire & Explosion Hazards.:
 Dangerous fire and explosive hazard.
 Vapor can travel distances to ignition source and flash back.

 6. ACCIDENTAL RELEASE MEASURES

Spill Response:
 Evacuate the area of all unnecessary personnel.
 Wear suitable protective equipment listed under Exposure /

       MSDS #XX0045   Page # 2 (continued on next page)

FIGURE 2 *(continued)*

Personal Protection.
Eliminate any ignition sources until the area is determined to be
free from explosion or fire hazards.
Contain the release and eliminate its source, if this can be done
without risk.
Take up and containerize for proper disposal as described under
Disposal.
Comply with Federal, State, and local regulations on reporting
releases. Refer to Regulatory Information for reportable
quantity and other regulatory data.
EM SCIENCE recommends Spill-X absorbent agents for various types
of spills. Additional information on the Spill-X products can be
provided through the EM SCIENCE Technical Service Department
(609) 354-9200.
The following EM SCIENCE Spill-X absorbent is recommended for
this product:

    SX0863   Solvent Spill Treatment Kit

7. HANDLING AND STORAGE

Handling & Storage:
  -Keep container closed and protected against physical damage
  Store in a cool, well-ventilated area, away from sources of ignition
  Do not breathe vapor.
  Do not get in eyes, on skin or on clothing
  Electrically ground all equipment when handling this product.
  Retained residue may make empty containers hazardous; use caution

8. EXPOSURE CONTROLS / PERSONAL PROTECTION

ENGINEERING CONTROLS AND PERSONAL PROTECTIVE EQUIPMENT:

Ventilation, Respiratory Protection, Protective Clothing, Eye Protection
  Material must be handled or transferred in an approved fume hood
  or with equivalent ventilation.
  Protective gloves (Viton, PVA or equivalent) should
  be worn to prevent skin contact
  Safety glasses with side shields should be worn at all times.
  NIOSH-approved respirator should be worn in the absence of
  adequate ventilation

Work / Hygenic Practices:
  Wash thoroughly after handling.
  Do not take internally.
  Eye wash and safety equipment should be readily available.

EXPOSURE GUIDELINES

FIGURE 2 *(continued)*

EM®

OSHA - PEL:

Component	TWA PPM	MG/M^3	STEL PPM	MG/M^3	CL PPM	MG/M^3	Skin
p-Xylene							

ACGIH - TLV:

Component	TWA PPM	MG/M^3	STEL PPM	MG/M^3	CL PPM	MG/M^3	Skin
p-Xylene	100	434	150	651			

9. PHYSICAL AND CHEMICAL PROPERTIES

Boiling Point (C 760 mmHg).: 138C
Melting Point (C)..........: 13C
Specific Gravity (H20 = 1).: 0.86
Vapor Pressure (mm Hg).....: 8.6   25C
Percent Volatile by Vol (%): 99.9+
Vapor Density (Air = 1)....: 3.7
Evaporation Rate (BuAc = 1): 0.7
Solubility in Water (%)....: Insoluble
Appearance.................:
    colorless liquid; aromatic odor

10. STABILITY AND REACTIVITY

Stability...............: Yes
Hazardous Polymerization:
    Does not occur

Hazardous Decomposition.:
    CO_x

Conditions To Avoid.....:
    Heat, sparks, open flame

Materials To Avoid......:
    ( )Water
    ( )Acids
    ( )Bases
    ( )Corrosives
    (X)Oxidizers
    ( )Other :

MSDS #XX0045   Page # 4 (continued on next page)

FIGURE 2   *(continued)*

EM®

11. TOXICOLOGICAL INFORMATION

Toxicity Data:

    orl-rat LD50: 8600 mg/kg   ihl-rat LC50: 4550 ppm/4H

Toxicological Findings:
    Tests on laboratory animals indicate material may produce adverse
    reproductive effects
    Cited in Registry of Toxic Effects of Substances (RTECS)

12. DISPOSAL CONSIDERATIONS

EPA Waste Numbers: D001   U239
Treatment:
    Incineration, fuels blending or recycle. Contact your local
    permitted waste disposal site (TSD) for permissible treatment
    sites.
    ALWAYS CONTACT A PERMITTED WASTE DISPOSER (TSD) TO ASSURE
    COMPLIANCE WITH ALL CURRENT LOCAL, STATE AND FEDERAL REGULATIONS.

13. TRANSPORT INFORMATION

DOT Shipping Name.........:
    p-Xylene

DOT Number...............:
    UN1307

14. REGULATORY INFORMATION

TSCA Inventory...........:
    The CAS number of this product is listed on the TSCA Inventory.

Component	SARA EHS (302)	SARA EHS TPQ (lbs)	CERCLA RQ (lbs)
p-Xylene			1000

	OSHA	SARA	DeMinimis

MSDS #XX0045   Page # 5 (continued on next page)

FIGURE 2   *(continued)*

EM®

Component	Floor List	313	for SARA 313 (%)
p-Xylene	Y	Y	1.0

15. OTHER INFORMATION

Comments:
    None
NFPA Hazard Ratings:
    Health   : 2
    Flammability : 3
    Reactivity   : 0
    Special Hazards:

Revision History:
    11/01/81 06/01/84 06/23/87 10/27/87 08/25/88

    | = Revised Section
    N/A = Not Available
    N/E = None Established

    The statements contained herein are offered for informational purposes
    only and are based upon technical data that EM SCIENCE believes to be
    accurate. It is intended for use only by persons having the necessary
    technical skill and at their own descretion and risk. Since conditions
    and manner of use are outside our control, we make NO WARRANTY, EXPRESS
    OR IMPLIED, OF MERCHANTABILITY, FITNESS OR OTHERWISE.

    Portions copyright Ariel Research Corporation, 1991.
    Restricted use conditions apply. Selected regulatory
    information in this MSDS has been derived from Ariel
    Research Corporation's International Chemical Regulatory
    Monitoring System (ICRMS). Use of this data is provided
    Subject to the terms of the License Agreement between EM
    Industries and Ariel Research Corporation. Further
    distribution is prohibited without authorization.

MSDS #XX0045   Page # 6

FIGURE 2 *(continued)*

B. Keeping the Notebook

If the pages in the notebook are not numbered, number them before using the book. Write your name and laboratory section number on the cover of your notebook. It is also advisable to put your address or telephone number on the book in case it is lost.

Leave two blank pages at the front of the book for a table of contents. Then, enter experiments consecutively in ink; use permanent ink, because your book will become splashed and stained with use.

Use only right-hand pages for writing up experiments. At the top of the page, write the date on which the experiment is performed. As you go along, leave plenty of space for notes that you might want to insert later. The empty left-hand pages may be used for calculations and jottings. If you will be running a distillation or determining more data for a particular experiment, be sure to leave blank pages as necessary before writing the procedure for the next experiment. If you make errors, do not rip out the page. Instead, line out errors (or draw an "X" over the entire page) and go on.

C. Entering Experiments

Each experiment in the notebook should contain the following information, along with any additional material required by your instructor.

1) Title
2) Balanced chemical equation, including other information described in the following paragraphs
3) Safety and disposal information
4) Procedure outline
5) Observations
6) Conclusions

These items are described in detail next.

1) The **title** of the experiment should briefly describe the experiment and should also contain the experiment number (and other reference, where appropriate). If the experimental objective is not clear from the title, a concise statement describing the experiment should be included.

2) The **balanced chemical equation** should show the formulas and names of the reactants and products. The molecular weight, the actual weight used, and the number of moles should be written under the name of each reactant. For your own convenience, pertinent physical properties, such as boiling points, should also be listed with the equation. Figure 3 shows a typical notebook page for the start of an experiment and also demonstrates the calculations you may have to perform.

The molecular weight and the **theoretical yield** should be placed under the name of each organic product. The theoretical yield of a product may be calculated from its molecular weight and the number of moles of the **limiting**

$$\underline{Experiment\ 10.1}$$

$$\underline{Synthesis\ of\ 1\text{-}Bromobutane\ from\ 1\text{-}Butanol}$$

$$CH_3CH_2CH_2CH_2OH + NaBr + H_2SO_4 \longrightarrow CH_3CH_2CH_2CH_2Br + NaHSO_4 + H_2O$$

1-butanol   sodium   sulfuric   1-bromobutane

             bromide   acid

<u>MW</u> : 74.12	102.90	98.08	137.03
<u>weight:</u> 18.5 g	30.0 g	25.0 mL (46.0 g)	34.2 g (theory)
<u>moles:</u> 0.250	0.292	0.469	0.250 (theory)

<u>calculation of numbers of moles of reactants:</u>

for 1-butanol: $\dfrac{18.5\ g}{74.12\ g/mol} = 0.250\ mol$

for NaBr: $\dfrac{30.0\ g}{102.90\ g/mol} = 0.292\ mol$

for H_2SO_4: $\dfrac{46\ g}{98.08\ g/mol} = 0.469\ mol$

<u>calculation of theoretical yield of product:</u>

for 1-bromobutane: $0.250\ mol \times 137.03\ g/mol = 34.2\ g$

FIGURE 3 A typical laboratory notebook page, including a balanced equation with the pertinent data, and calculations of moles of reactants and theoretical yield of product.

reagent (the reactant present in shortest supply). In the example in Figure 3, NaBr and H_2SO_4 are present in excess. Therefore, 1-butanol is the limiting reagent. In the example, the maximum number of moles of product that could be obtained from 0.250 mol of 1-butanol is 0.250 mol, or 34.2 g, as shown in the last calculation in Figure 3. Appendix I describes yield calculations in more detail.

If the experiment is not a reaction but is an isolation or purification experiment, then the formulas, names, and so on, of the compounds in question should be written out.

3) **Safety and disposal information** can be obtained from MSDSs provided to you by your instructor. You should verify that your disposal procedures are in accordance with the rules established for your laboratory.

crude reaction mixture:

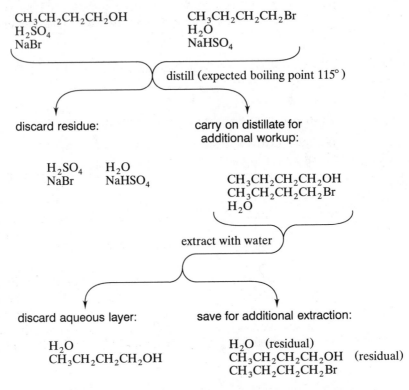

FIGURE 4 A partial flow diagram for the isolation of 1-bromobutane.

4) The **actual procedure** that will be used should be outlined *in detail*. The purpose of the outline is to provide an overview of what you will be doing in the laboratory, thus allowing you to organize your time efficiently. (Do not waste your time by copying the procedure word for word. You can always refer back to the original procedure if necessary.) Boiling points of solvents and any special hazards, such as flammability, should also be noted in your outline.

A **flow diagram,** showing how the product will be isolated from by-products or unreacted starting material, must be included. Figure 4 shows a portion of such a diagram.

5) **Observations** should be recorded in your notebook *as you do the experiment*. Examples of observations might be "The ethanol solution was yellow," or "Approx. 5 mL of the reaction mixture was spilled and lost while being transferred to a separatory funnel." It is better to include too many observations than too few; however, use common sense. Entering the fact that you went to the storeroom to get a beaker wastes both your time and the time of anyone who reads your notebook. On the other hand, failure to record a weight or a physical constant may also waste your time—you may have to repeat the experiment.

6) **Conclusions** of your experiment will usually include the weight and physical properties (such as melting point or boiling point) of the isolated and purified product plus the **percent yield**. This percent yield is the actual percent of the theoretical yield that you obtained.

$$\text{percent yield} = \frac{\text{actual yield in g}}{\text{theoretical yield in g}} \times 100$$

Thus, you might write in your notebook

yield: 5.3 g (61%) of benzoic acid, mp 117°–119°.
mp of an authentic sample of benzoic acid, 119.5°–120°.
mixed mp of product with authentic sample, 118.5°–119°.

In some experiments, you may have to prove the structure of a compound. In these cases, your conclusion should include all the supporting data used in the structure proof. These data may consist of physical constants, spectroscopic information, and chemical reactivity.

You may be required to turn in reaction products to your instructor. Your instructor will provide sample bottles for solids and for liquids or will suggest other containers. Any sample you turn in should be labeled clearly with the compound's name, weight, and pertinent physical constants. Your name and laboratory section number should also be included. Although each instructor has a preferred label format, a typical label looks like the following.

Benzoic acid
mp 117° - 119°
5.3g (61% yield)
Expt. 1b-1, Lab section 28
Jane Doe

Use waterproof ink for your label. Your instructor may also want you to cover the label with transparent tape.

3 Laboratory Equipment

A. Glassware

Figure 5 shows the glassware found in a typical student locker. Your own laboratory may not supply every item shown or it might contain items not shown.

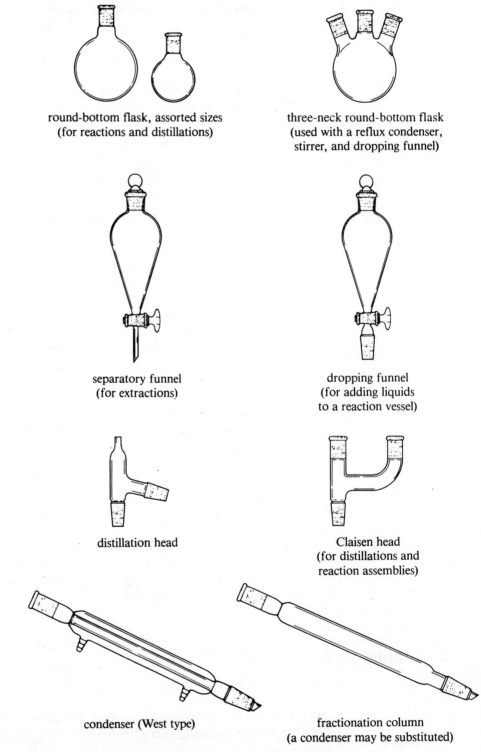

round-bottom flask, assorted sizes
(for reactions and distillations)

three-neck round-bottom flask
(used with a reflux condenser,
stirrer, and dropping funnel)

separatory funnel
(for extractions)

dropping funnel
(for adding liquids
to a reaction vessel)

distillation head

Claisen head
(for distillations and
reaction assemblies)

condenser (West type)

fractionation column
(a condenser may be substituted)

FIGURE 5   Typical glassware stored in a student locker.

dropper
(disposable pipet)

Erlenmeyer flask,
assorted sizes

heavy-walled filter flask
(for vacuum filtration)

beaker, assorted sizes

funnel (conical)

powder funnel
(with thick stem)

stemless funnel

Büchner funnel
(for vacuum filtration)

Hirsch funnel
(for vacuum filtration)

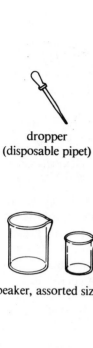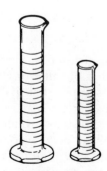

graduated cylinder,
assorted sizes

distillation adapter

vacuum distillation
adapter

drying tube

watch glass

FIGURE 5 (continued)

Note the ground-glass joints on the round-bottom flask, condenser, and so on. These joints are ground to a *standard taper* (⊥). The size of a standard-taper joint is identified by a pair of numbers, such as ⊥19/22, ⊥14/20, or ⊥24/40. In each pair, the first number refers to the outside diameter of the inner joint at its widest part, and the second number refers to the length of the joint. Any 19/22 inner joint will fit any 19/22 outer joint; therefore, glassware of the same standard taper is interchangeable. The advantages of ground-glass joints are that they provide a good seal between two pieces of equipment and that a laboratory setup can be assembled quickly.

One disadvantage of ground-glass joints is their tendency to stick, or freeze. A film of hydrocarbon or silicone grease on these joints is used to prevent sticking. When you assemble ground-glass equipment, grease each inner joint *lightly*, insert it into the outer joint, and rotate the joints to distribute the grease evenly. The glass stopcocks of separatory funnels or dropping funnels also require grease, but Teflon® stopcocks do not. The cleaning and storing of ground-glass equipment is discussed in Section 4.

Equipment with ground-glass joints is expensive. If you break a piece of glassware containing a ground-glass joint, do not discard the joint unless you are told to do so by your instructor. A competent glassblower can recycle the joints by sealing them on ordinary, less expensive flasks and condensers.

If your laboratory does not have ground-glass equipment, you must use corks and rubber stoppers. Your instructor will provide instructions for boring holes in corks and stoppers.

When you check the equipment in your locker, check each piece of glassware carefully for cracks and "stars" (small star-shaped cracks). Each time you use a piece of glassware, recheck it. Any cracked or starred glassware should be returned to the storeroom and replaced by undamaged ware. In some cases, a glassblower will be able to repair the cracks.

B. Nonglass Equipment

Other useful items often found in a student locker are pictured in Figure 6. The Filtervac and neoprene adapters are used to attach a Büchner or Hirsch funnel to a filter flask. However, a one-holed rubber stopper that fits both funnel and flask gives a better seal and is easier to use.

Polyethylene wash bottles are useful for holding distilled water and most solvents. If yours is the type with a small hole in the neck, place your finger over the hole and squeeze the bottle to force liquid through the tip of the tube. Release your finger from the hole to release the pressure and stop the flow of liquid.

C. Heating Equipment

Several devices for heating liquids in flasks are shown in Figure 7. The **steam bath** heats liquids to a maximum of about 90° and is useful for heating low-boiling solvents, especially flammable ones. To clear water from the steam line, turn the steam on forcefully. Then turn the steam to low before placing a flask on it. (Too much steam will allow water to contaminate the contents of the flask, and its noise is irritating to others in the laboratory.) A round-bottom flask can be set into the appropriately sized ring of the steam bath, while an Erlenmeyer flask can be set on top of a ring smaller than itself. If all the rings are left on top of the steam bath, two or three small Erlenmeyer flasks can be warmed simultaneously. Many modern **hot plates** are also safe for heating flammable solvents.

SAFETY NOTE Some hot plates develop very hot surfaces, have exposed electrical coils, or spark when the thermostat clicks on and off. These hot plates should *not* be used with flammable solvents because of the danger of fire or explosion.

A **heating mantle** is used for heating the contents of a round-bottom flask. With a heating mantle, the flask becomes hotter than its contents. To avoid decomposition of material splashed on the hot glass, never use a heating mantle with a nearly empty flask.

Heating mantles are available in assorted sizes. A soft, glass coil mantle should fit a flask snugly. A hard, ceramic coil mantle need not fit a flask snugly and can thus accommodate more than one flask size.

SAFETY NOTE If the heating mantle in your locker appears worn or if the heating element is exposed, return it to the storeroom. A worn heating mantle can cause a fire!

The electrical plug of a typical heating mantle will not fit into the standard wall socket, because heating mantles are not designed to operate at 110 volts. Instead, the mantle is plugged into a **variable transformer**, or **rheostat** (Variac®, Powerstat®, Powermite®), which is used to adjust the voltage and thus the temperature. Figure 7 shows a transformer of this type.

In some laboratories, **heat lamps** are used as heat sources. The principal disadvantage of heat lamps is that they are easily broken. Breakage can occur, for example, if water or other liquid splashes on a hot bulb.

Bunsen burners and microburners are used for bending glass tubing and for other glassworking. They may also be used to heat aqueous solutions and high-boiling liquids. The use of a wire gauze with a ceramic center is

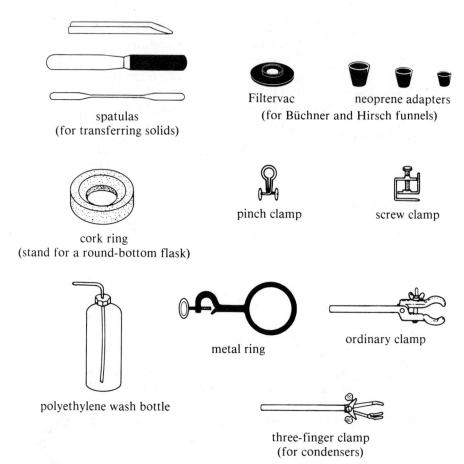

FIGURE 6 Hardware and other nonglass items typically found in a student locker.

recommended to distribute the heat from a burner to a flask and to prevent localized overheating.

SAFETY NOTE Burners should be used with the utmost caution in the organic laboratory because of the ever-present vapors of organic solvents. *NEVER* light a burner (or even a match) if someone is using a flammable solvent nearby. (Check with your instructor if you are unsure whether it is safe to light a match.) *NEVER* use a burner to heat a flammable solvent. *NEVER* open a solvent bottle without first checking the vicinity for flames. Heat flammable solvents only in the hood.

• •

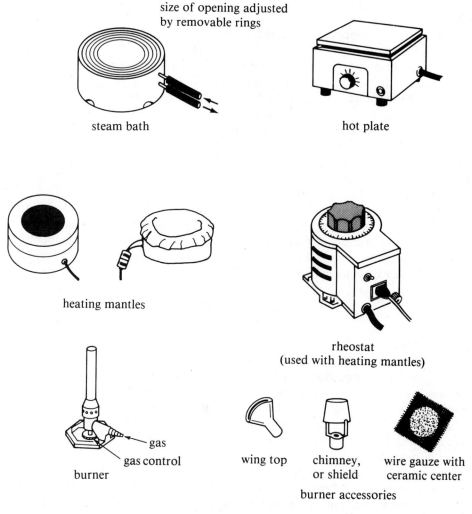

size of opening adjusted
by removable rings

steam bath

hot plate

heating mantles

rheostat
(used with heating mantles)

gas
gas control
burner

wing top

chimney,
or shield

wire gauze with
ceramic center

burner accessories

FIGURE 7 Some typical laboratory heating devices.

D. Community Equipment

You will share a **balance** with your classmates. Your instructor will demonstrate the operation of the balance or balances in your laboratory. A balance is a delicate instrument: treat it with care, and always wipe up spills (liquid or solid) on or near the balance. Do *not* blow spilled powders off the balance pan—many chemicals that are ordinarily safe to handle are dangerous when inhaled. When you finish with a balance, return it to a weight of zero if this is not accomplished automatically. This will help prevent uneven wearing of the weighing mechanism.

You will probably be using analytical electronic equipment such as gas chromatographs and spectrometers in this course. These are also delicate instruments. Do not be a knob twister. Understand and follow your instructor's directions.

If you encounter problems with any of the laboratory's electrical or electronic equipment, notify your instructor.

The **fume hood** should be operating whenever someone is working in the laboratory. It is recommended that highly flammable or toxic materials *always* be handled in the hood. Unfortunately, space limitations in a student laboratory do not always permit this. Because of the nature of the chemicals used in the hood, it is imperative that the bench in the hood be kept clean. If you find some spilled chemical of unknown origin on the hood bench, clean it up as if it were toxic, corrosive, and flammable.

The sharing of other community equipment is a matter of safety and courtesy. Keep reagent bottles tightly capped. Return them to their proper spots promptly and do not take them to your workstation. Do not contaminate chemicals in reagent bottles. If you see that a bottle is almost empty, report the fact to your instructor.

· · · · · · · · · · ·
4 Cleaning Glassware

Glassware for organic chemical reactions should be both *clean* and *dry*. The presence of water in a flask can ruin many experiments (except, of course, those performed in aqueous solution). For this reason, glassware should be washed as soon as you finish using it (certainly by the end of each laboratory period) and then allowed to drain and dry in your locker until the next laboratory period.

Another reason to wash your glassware promptly is that freshly dirtied glassware is easier to clean than glassware containing dried-out tars and gums. Furthermore, some compounds, like sodium hydroxide and potassium hydroxide, can etch glass and ruin ground-glass joints if left standing in flasks.

A. General Cleaning

Most glassware can be cleaned readily with strong industrial detergent or scouring powder and a bottle brush. The brush should be bent, if necessary, to reach the entire inside of a flask. The detergent should be rinsed out thoroughly, then the piece of glassware rinsed with a small amount of distilled water. A flask can be stored upside down on a crumpled towel in a beaker to drain and dry. Other pieces of equipment, like condensers, can be laid on their sides in the locker.

When absolutely necessary, glassware can be dried quickly by rinsing it with 5–10 mL of *acetone*, $(CH_3)_2C=O$. (Because of the expense, some laboratories do not allow acetone to be used for rinsing flasks.) Never use expensive reagent-grade acetone for rinsing flasks! Use "wash acetone" instead. If it is used to rinse only water from a flask, wash acetone may be reused several times before it is discarded in the waste acetone container. Acetone, like any solvent, should be handled with care. It should not be poured on the skin, and its vapors should not be inhaled.

A flask rinsed in acetone will dry fairly quickly in air. The drying process can be speeded up by placing a clean glass dropper, connected to a vacuum line by heavy-walled rubber tubing, in the drained flask. The vacuum will suck fresh air through the flask, sweeping acetone and water vapors into the vacuum line. Using a stream of air to dry an acetone-rinsed flask is not recommended. Compressed air is likely to contain droplets of water and oil that will contaminate the flask. Also, a noisy blast of air shot into a flask can startle other people working in the laboratory.

Never use a flame to dry a flask. If the flask contains water droplets, the flask may become unevenly heated and crack. If the flask contains solvents, the vapors may catch fire. A drying oven may be used to dry water-rinsed glasswater but not glassware rinsed with a solvent. Before placing glassware in the oven, separate any ground-glass joints and remove stopcocks. Do not dry Teflon® stopcocks in a drying oven.

Ground-glass equipment should be dismantled and cleaned before it is placed in your locker. The joints should be kept free of chemicals and grit; otherwise, they may become stuck together, or frozen. A frozen joint can sometimes be unfrozen by rinsing the outer portion in hot water to cause it to expand. A *gentle* rap on the tabletop might also loosen it. There are some more sophisticated techniques for unfreezing joints; your instructor can advise you.

Glass stopcocks in separatory funnels are treated as other ground-glass joints and should be cleaned of hydrocarbon greases with a tissue dampened with acetone and stored separately. (Silicone greases should be carefully removed with dichloromethane.) If you would rather keep your stopcock in the separatory funnel, regrease it before replacing it. Teflon stopcocks, which do not need greasing, should be cleaned and then replaced loosely in the joint until they are used again.

B. Hard-to-Clean Flasks

Tars and gums, which are large organic molecules called **polymers**, are formed when a large number of organic molecules react to yield very long chains or three-dimensional molecular networks. These substances are not soluble in water. Therefore, they should be scraped out of a flask with a metal spatula and discarded in a waste crock, not in the sink. Acetone is a good solvent for most organic compounds and is often useful for dissolving remnants of tars and other organic residues from dirty flasks. For these cleaning purposes, *waste acetone*, impure acetone that is to be discarded, should be used. In some cases, swirling 5–10 mL of acetone in a flask will dissolve the organic residues. In other cases, several hours of soaking will be necessary. If you encounter such a stubborn tar, cork the dirty flask containing a small amount of acetone and store it upright in your locker until the next laboratory period.

Suggested Readings*

Steere, N. V. *Handbook of Laboratory Safety*. 2nd ed. Cleveland, Ohio: CRC Press, 1971.

Manufacturing Chemists' Association (now the Chemical Manufacturing Association). *Guide for Safety in the Chemical Laboratory*. 2nd ed. New York: Van Nostrand Reinhold, 1972.

Green, M. E., and Turk, A. *Safety in Working with Chemicals*. New York: Macmillan, 1978.

Safety in Academic Chemistry Laboratories. 4th ed. Washington, D. C.: American Chemical Society, 1985.

Reese, K. M. *Health and Safety Guidelines for Chemistry Teachers*. Washington, D. C.: American Chemical Society, 1979.

National Research Council. *Prudent Practices for Handling Hazardous Chemicals*. Washington, D. C.: National Academy Press, 1981.

Fawcett, H. H., and Wood, W. S., eds. *Safety and Accident Prevention in Chemical Operations*. New York: Wiley, 1982.

Bretherick, L., ed. *Hazards in the Chemical Laboratory*. 4th ed. London: The Royal Society of Chemistry, 1986.

Problems

1   Give reasons for the following safety rules.
   (a) Contact lenses should not be worn in the laboratory.
   (b) A chemical spill on the skin should be washed off with water, not with solvent.
   (c) Solvents are not to be poured into the drainage trough.
   (d) Water should not be used to extinguish laboratory fires.
   (e) To dilute concentrated sulfuric acid, we pour it onto ice instead of simply mixing it with water.

*See also the toxicology references in Appendix III and the waste disposal references in Appendix IV.

2   What should you do in each of the following circumstances?
(a) Your neighbor splashes a chemical into his or her eye.
(b) A strong acid spills onto your hands.
(c) Your neighbor's clothing catches fire.
(d) Your neighbor's flask catches fire.

3   What are the principal hazards of each of the following solvents?
(a) Carbon disulfide   (b) Diethyl ether
(c) Benzene   (d) Carbon tetrachloride

4   Make the following conversions:
(a) 5.0 g $CH_3CH_2CH_2CH_2CH_2Br$ to moles
(b) 10.0 mL concd H_2SO_4 (96%, density 1.84) to moles
(c) 0.100 mol CH_3OH to grams
(d) 2.50 mol NaBr to grams
(e) 0.30 mol H_2SO_4 to mL of $6N$ H_2SO_4

5   Calculate the percent yield when (a) the theoretical yield of a product is 15.3 g and a student obtains 6.9 g; (b) the theoretical yield is 3.1 g and a student obtains 2.7 g.

6   For each of the following reactions, (1) identify the limiting reagent, and (2) calculate the theoretical yield of the organic product. (*Note:* The equations as shown are not necessarily balanced.)
(a) $CH_3CO_2H + NaOH \longrightarrow CH_3CO_2Na + H_2O$
      25.0 g   10.0 g
(b) $H_2NCH_2CH_2CH_2NH_2 + HCl(12M) \longrightarrow Cl^- H_3\overset{+}{N}CH_2CH_2CH_2\overset{+}{N}H_3Cl^-$
          5.0 g   10.0 mL

7   What would be the heat source (or heat sources) of choice for boiling each of the following solvents?
(a) Diethyl ether, $(CH_3CH_2)_2O$, bp 35°
(b) Water, bp 100°
(c) Ethanol, CH_3CH_2OH, bp 78°
(d) Acetone, $(CH_3)_2C{=}O$, bp 56°

TECHNIQUE 1

Crystallization
• •

When a solid organic compound is prepared in the laboratory or isolated from some natural source, such as leaves, it is almost always impure. A simple technique for the purification of such a solid compound is **crystallization.** To carry out a crystallization, dissolve the compound in a minimum amount of hot solvent. If insoluble impurities are present, the hot solution is filtered. If the solution is contaminated with colored impurities, it may be treated with decolorizing charcoal and filtered. The hot, saturated solution is finally allowed to cool slowly so that the desired compound crystallizes at a moderate rate. When the crystals are fully formed, they are isolated from the **mother liquor** (the solution) by filtration.

If an extremely pure compound is desired, the filtered crystals may be subjected to **recrystallization.** Of course, each crystallization results in some loss of the desired compound, which remains dissolved in the mother liquor along with the impurities.

Crystallization is not the same as precipitation. Precipitation is the *rapid* formation of an amorphous solid while crystallization is the *slow* formation of a crystalline solid. If a hot, saturated solution is cooled too quickly, the compound may precipitate instead of crystallizing. A precipitated solute may contain many impurities trapped in the rapidly formed amorphous mass by entrainment. On the other hand, when a solution is allowed to crystallize slowly, impurities tend to be excluded from the growing crystal structure, because the molecules in the crystal lattice are in equilibrium with the molecules in solution. Molecules unsuitable for the crystal lattice are likely to remain in the solution, and only the most suitable molecules are retained in the crystal structure. Because impurities are usually present in low concentration, they remain in solution even when the solution cools.

To understand why a slow and careful crystallization is preferable to a rapid precipitation, consider the mechanism of crystallization. Crystallization occurs in stages. As the hot, saturated solution cools, it becomes supersaturated; and then crystal nuclei form. These nuclei often form on the walls of the container, at the liquid surface, or on a foreign body (such as a dust particle), because there is a greater probability of proper molecular association at these locations.

Once the crystal nuclei have been formed, additional molecules migrate to their surfaces by diffusion and join the crystal lattice. Because the molecules must migrate from the bulk of the solution to the growing crystal surface, the solution surrounding the crystal becomes less concentrated than the bulk of the solution. Also, crystal growth is usually exothermic. So the heat released from the growing crystal increases the solubility of the compound near the surface. For crystallization to continue, the concentration of solute at the crystal site must be increased and the heat must be dissipated. These processes occur by diffusion and take time. Premature chilling or agitation can increase the rate of crystal growth to the point at which the solid precipitates. The purest crystals are obtained when crystallization occurs slowly from an undisturbed solution.

· · · · · · · · · ·
1.1 Solvents for Crystallization

The ideal solvent for the crystallization of a particular compound is one that (1) does not react with the compound; (2) boils at a temperature below the compound's melting point; (3) dissolves a moderately large amount of the compound when hot; (4) dissolves only a small amount of compound when cool; (5) is moderately volatile so that the final crystals can be dried readily; and (6) is nontoxic, nonflammable, and inexpensive. In addition, impurities should be either highly insoluble in the solvent (so that they can be filtered from the hot solution) or else highly soluble (so that they remain in solution during the crystallization). As you might guess, a solvent possessing *all* these attributes does not exist.

The primary consideration in choosing a solvent for crystallizing a compound is that the compound be moderately soluble in the hot solvent and less so in the cold solvent. Unfortunately, the solubility of a compound in a solvent cannot be predicted with accuracy. Most commonly, in the selection of a specific solvent for a specific compound, the solubility of the compound in various solvents is determined by trial and error.

General guidelines for predicting solubilities based upon the structures of organic compounds do exist. For example, an *alcohol*, a compound containing the hydroxyl ($-OH$) group as its functional group, may be soluble in water because it can form hydrogen bonds with water molecules. *Carboxylic acids* (compounds containing $-CO_2H$ groups) and amines (compounds containing $-NH_2$, $\rangle NH$, or $\rightarrow N$ groups) also can form hydrogen bonds and are also generally soluble in polar solvents such as water or alcohols.

As the amount of hydrocarbon in the compound increases, the compound's solubility in water will decrease; but it still may be soluble in an alcohol, such as ethanol. Compounds that are largely hydrocarbon in structure are not soluble in polar solvents because C—C and C—H bonds are not polar. For these compounds, we would choose a nonpolar solvent—for example, low-boiling petroleum ether, which is a mixture of alkanes such as pentane, $CH_3(CH_2)_3CH_3$, and hexane, $CH_3(CH_2)_4CH_3$. Thus, in choosing crystallization solvents, chemists generally follow the rule of the thumb: **like dissolves like.**

If the best solvent for crystallizing a compound is not known, small portions of the compound can be tested with a variety of likely solvents (see Section 1.3A). For well-known compounds, however, suitable crystallization solvents have already been determined. This information can be found in textbooks, handbooks, and chemical journals. (The *Handbook of Chemistry and*

TABLE 1.1   Some common crystallization solvents

Name	Formula	ϵ^a	bp (°C)	Miscibility with water	Comments
water	H_2O	78.5	100	—	—
methanol	CH_3OH	32.6	65	yes	flammable, toxic
ethanol (95%)	CH_3CH_2OH	24.3	78	yes	flammable
acetone (propanone)	$(CH_3)_2C=O$	20.7	56	yes	flammable
methylene chloride (dichloromethane)	CH_2Cl_2	9.1	41	no	—
ethyl acetate	$CH_3CO_2CH_2CH_3$	6.0	77	no	flammable
chloroform (trichloromethane)	$CHCl_3$	4.8	61	no	toxic
diethyl ether (ethyl ether, ether)	$(CH_3CH_2)_2O$	4.3	35	no	highly flammable
toluene	$C_6H_5CH_3$	2.4	111	no	flammable
benzene	C_6H_6	2.3	80	no	flammable, toxic, carcinogenic, freezes at 5°
carbon tetrachloride (tetrachloromethane)	CCl_4	2.2	77	no	toxic
cyclohexane	C_6H_{12}	2.0	81	no	flammable, freezes at 6.5°
ligroin[b]	C_nH_{2n+2}	~ 1.9	—	no	flammable
petroleum ether[c]	C_nH_{2n+2}	~ 1.8	30–60	no	flammable

[a]Dielectric constant (a measure of polarity) at about 20°–25°. The solvents are shown in order of decreasing polarity.
[b]A mixture of alkanes boiling at 60°–90°, 90°–150°, or other specified temperature range. Ligroin is occasionally referred to as "high-boiling petroleum ether."
[c]A mixture of alkanes and not a true ether.

TABLE 1.2 Some common solvent pairs for crystallization

methanol–water	diethyl ether–methanol
ethanol–water	diethyl ether–acetone
acetone–water	diethyl ether–petroleum ether
benzene–ligroin	methanol–dichloromethane

Physics, CRC Press, Inc., Boca Raton, Florida, lists some solubility data.) Table 1.1 lists some common crystallization solvents, arranged according to their polarities.

Ideally, a compound to be crystallized should be soluble in the hot solvent but insoluble in the cold solvent. When such a solvent cannot be found, a chemist may use a **solvent pair.** A solvent pair is simply two miscible liquids chosen so that one liquid dissolves the compound readily and the other does not. For example, many polar organic compounds are very soluble in ethanol but insoluble in water. To crystallize such a compound, dissolve it in a moderate amount of hot ethanol; then add water drop by drop until the solution becomes turbid (cloudy). Finally, add a few drops of ethanol to redissolve the precipitating compound. The resulting ethanol–water solution is a saturated solution and is allowed to cool slowly so that crystallization will occur. Table 1.2 lists some common solvent pairs.

1.2 Steps in Crystallization

1) Dissolving the compound. The first step in crystallization is dissolving the compound in a *minimum amount* of the appropriate hot solvent in an Erlenmeyer flask. An Erlenmeyer flask is used instead of a beaker or other container for several reasons. The solution is less likely to splash out and dust is less likely to get in. The sloping sides allow boiling solvent to condense and return to the solution and allow easy removal of crystals. Also, an Erlenmeyer flask can be corked and stored in your locker.

Boiling chips, or **boiling stones,** should be used whenever a solvent is brought to a boil, unless the liquid can be *constantly* stirred or swirled. Boiling chips are small porous stones of calcium carbonate or silicon carbide that contain trapped air. When the chips are heated in a solvent, they release tiny air bubbles, which ensure even boiling. When all the air has been released, the stone's pores provide cavities where bubbles of solvent vapor can form. Without boiling chips, part of the solvent may become superheated and boil in spurts, a process called **bumping.** Bumping is likely to cause some of the solution to splash out of the flask. Only two or three boiling chips are necessary to prevent bumping.

Boiling chips should never be added to a hot solution! If a solution is at or near its boiling point when boiling chips are added, the solution will almost certainly boil out of the flask.

Boiling chips will maintain their function throughout a long boiling period. But once they are used and cooled, the pores fill with liquid and they lose their ability to release bubbles. Therefore, use fresh boiling chips each time you heat a solution.

Boiling sticks may be used instead of boiling chips. Propped upright in a flask, these small wooden sticks provide a porous surface on which the solvent bubbles can form. As with boiling chips, a fresh stick must be used each time the solution is heated.

Pulverize a lumpy solid with a spatula to dissolve it more rapidly. To ensure that a minimum amount of solvent is used, add the solvent a few milliliters at a time and heat the mixture with constant stirring or swirling. When almost all of the solid has dissolved, examine the solution and the bottom of the flask for insoluble impurities. If impurities are visible, do not add excess solvent in an attempt to dissolve them, but filter the hot solution. This filtration is not necessary and is, in fact, undesirable if the solution looks clear and clean. If the solution appears to be contaminated with colored impurities, decolorizing charcoal may be added at this time. The use of decolorizing charcoal is discussed in Section 1.3B.

SAFETY NOTE   Toxic solvents should be heated only in a hood. A Bunsen burner should be used only for aqueous solutions — and only when no flammable solvents are being used in the vicinity. Some hot-plate surfaces are also capable of igniting flammable solvents. The safest source for heating a solvent, especially a low-boiling solvent, is a steam bath. Section 3C of the introduction describes its use.

2) Filtering insoluble impurities.   Filtering a hot, saturated solution inevitably results in cooling and in evaporation of some of the solvent. Therefore, a premature crystallization of the compound on the filter and in the funnel may occur. A few precautions can minimize this premature crystallization.

To help prevent clogging in the funnel, choose a stemless funnel, a short-stemmed funnel, or a powder funnel. Preheat the funnel by placing boiling chips and a small amount of solvent in the receiving Erlenmeyer flask, resting the funnel on top, and heating. Alternatively, warm the funnel on the flask containing the hot solution to be filtered.

Before filtering, add a little extra solvent (about 5–10% of the total volume) to the solution, and keep the solution hot while preparing the filtration apparatus. Filter the hot solution through either a plug of glass

Step 1

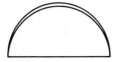

Step 2

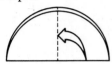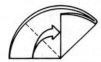

Step 3

fold to the back

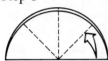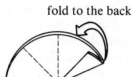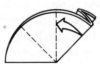

fold back

fold back

Step 4

FIGURE 1.1 How to prepare fluted filter paper.

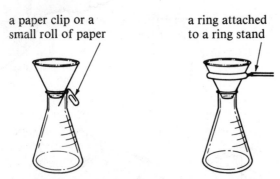

a paper clip or a
small roll of paper

a ring attached
to a ring stand

FIGURE 1.2 Ways to create an air space between a funnel and an Erlenmeyer flask.

wool or filter paper. Filter paper is rated by its porosity. For hot filtration, a
porous paper, such as Whatman's No. 1 or No. 4, should be used. (Do not use
Whatman's No. 5 or No. 6 papers, which have slow filter speeds.) **Fluted
filter paper** is preferred to folded filter paper, because the increased surface
area of the fluted paper allows the filtration to proceed more rapidly. Figure
1.1 shows how to prepare a piece of fluted filter paper. Place the fluted filter
paper in the warm funnel in the neck of the receiving flask. Figure 1.2 shows
two ways that the funnel can be supported slightly away from the lip of the
flask in order to prevent a liquid seal from blocking the flow of air and
solvent fumes.

To pour the hot solution, wrap the hot flask in a towel or hold it in a
clamp. Do not use a test tube clamp or tongs, because they do not have
enough strength to hold the flask. Alternatively, use a pair of inexpensive
cotton gloves.

During the filtration, keep both flasks hot on a steam bath or hot plate.
To keep the solution hot, pour only small amounts into the filter paper
(instead of filling the filter paper to the brim). If a flammable solvent and a
hot plate are used, move the flasks away from the hot plate when pouring so
that solvent vapors do not flow over the heating element.

If crystallization occurs in the funnel, you can often remove the crystals
by heating the receiving flask to boiling with the funnel still on it. Solvent
condensing in the funnel may dissolve the crystals and carry them back to
the filtered solution. Alternatively, wash the solid into the flask with a little
hot solvent.

After all of the hot solution has been filtered, wash the original flask
with a small amount of hot solvent. Pour this solvent through the filter paper
into the receiving flask to transfer the final traces of the desired compound.
Two washings may be necessary; however, use a minimum amount of
solvent.

The crystallization flask should contain a hot, clear, saturated solution
of the compound. Boil away excess solvent at this time using boiling chips or

a boiling stick to prevent bumping. (Remember to use the hood for a toxic or flammable solvent.) If the hot solution starts to crystallize, reheat it to dissolve the crystals. If so much solvent has evaporated that these crystals will not redissolve, add a small additional amount of solvent to the flask and bring the mixture to a boil.

3) Crystallizing the compound. Cover the flask containing the hot, saturated solution with a watch glass or inverted beaker to prevent solvent evaporation and dust contamination. Then, set the flask aside where it can remain undisturbed (no jostling or bumping, which will induce precipitation rather than crystallization) for an hour or several hours. If the flask must sit for several days, allow it to cool to room temperature. Then stopper it with a cork (not a rubber stopper if an organic solvent was used) to prevent solvent evaporation.

Chilling the mixture in an ice–water bath after crystallization appears complete will increase the yield of crystals. Be sure to allow ample time for the final crystal growth to occur before chilling.

Sometimes a hot solution cools to room temperature with no crystallization occurring. In such a case, your first question should be, "Is the solution *supersaturated*?" Often, crystallization can be induced in a supersaturated solution by scratching the inside of the flask up and down at the surface of the solution with a glass rod. The scratching of the glass is thought to release microcrystals of glass, which serve as a template, or seeds, for crystal growth. If scratching the flask does not start the crystallization, a **seed crystal** may be added. A seed crystal is a small crystal of the original material set aside to provide a nucleus upon which other crystals can grow. Sometimes seed crystals can be obtained from the glass rod used for scratching, after the solvent has evaporated from it. Allowing a few drops of solution to evaporate on a watch glass may also produce seed crystals. After addition of the seed crystal, set the flask aside to allow for crystallization to proceed.

If scratching and seeding do not produce crystals, your next question should be, "Did I use too much solvent?" If more than the minimum amount of solvent was used in the earlier steps, the excess must be boiled away and the flask again set aside to crystallize. Unless the solid begins to separate again after evaporation, reduce the volume of the solution by one-third. This is usually sufficient to produce a saturated solution.

Another problem encountered in crystallization is **oiling out:** instead of crystals appearing, an oily liquid separates from solution. A compound may oil out if its melting point is lower than the boiling point of the solvent. A very impure compound may oil out because the impurities depress its melting point. The formation of an oil is not selective, as is crystallization; therefore, the oil (even if it solidifies) is probably not a pure compound.

Reheat a mixture that has oiled out in order to dissolve the oil (add more solvent if necessary). Then allow the solution to cool slowly, perhaps adding a seed crystal or scratching with a glass rod. If the substance has a low melting point, a lower-boiling solvent may be necessary. Alternatively, use more solvent and keep the temperature of the solvent below the melting point of the solute.

If these techniques do not prevent oiling out, allow the oil to solidify (a seed crystal or chilling may be necessary), filter the solid or decant (pour off) the solvent, and crystallize the solid using fresh or a different solvent. Enough impurities may have been removed in the attempted crystallization that the second one will proceed smoothly.

4) Isolating the crystals.   Crystals are separated from their mother liquor by filtration. Gravity filtration (simple filtration using fluted filter paper and a funnel) should be used if the crystallization solvent is a low-boiling organic solvent. Vacuum filtration (see below) should not be used, because the solvent will evaporate and contaminate the crystals with the impurities that have just been removed.

Vacuum filtration apparatus   If water or a high-boiling organic solvent has been used, then **vacuum** or **suction filtration** is the procedure of choice. Vacuum filtration has the advantage of being much faster than gravity filtration. It has the disadvantage of requiring more equipment. Figure 1.3 shows the physical setup required for vacuum filtration. The trap is neces-

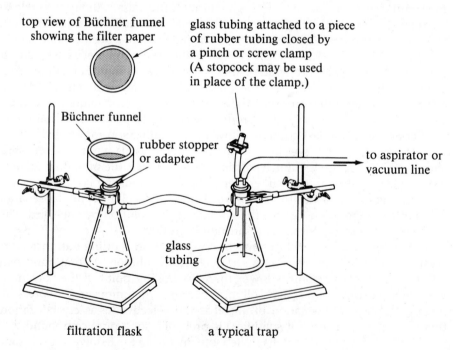

top view of Büchner funnel
showing the filter paper

glass tubing attached to a piece
of rubber tubing closed by
a pinch or screw clamp
(A stopcock may be used
in place of the clamp.)

Büchner funnel

rubber stopper
or adapter

to aspirator or
vacuum line

glass
tubing

filtration flask

a typical trap

Note: Flasks and tubing are heavy-walled.

FIGURE 1.3   A vacuum filtration apparatus.

sary regardless of whether a water aspirator or a centralized vacuum system is used. The purpose of this trap is to (1) prevent any solution from being accidentally sucked into the vacuum line and (2) prevent any water from an aspirator from backing up into the filter flask. With a trap, this water will be caught before it contaminates the mother liquor.

Heavy-walled vacuum tubing must be used for vacuum connections, because ordinary tubing collapses when vacuum is applied. All flasks should be clamped to ring stands. The filter flask, especially, should be firmly clamped, because it usually becomes top-heavy when a Büchner funnel and vacuum line are connected to it.

Attach a Büchner funnel or Hirsch funnel to the filter flask with a rubber adapter or a one-holed rubber stopper (best) so that the connection will be airtight when vacuum is applied. Place a medium- or slow-speed filter paper (such as Whatman's No. 2, 5, or 6) on the perforated surface of the funnel. (A fast-speed, porous filter paper allows finely divided solids to pass through under vacuum.) The filter paper must lie flat and not curl up at the sides, yet it must cover all the holes. When the vacuum is applied, the filter paper is pulled snugly to the flat surface of the funnel by suction. To ensure no leakage around the edges, moisten the filter paper with the solvent before applying vacuum.

Water aspirator   Many laboratories are equipped with **water aspirators,** devices that attach to faucets and develop a vacuum through a side tube when water flows through the main tube. Place a large beaker in the sink under the water outlet to minimize splashing. Because aspirators are easily plugged, they should be checked before each use. Turn the water on *full force* and hold your finger on the vacuum hole to feel the suction before attaching the rubber tube of your filtration apparatus.

As previously mentioned, an aspirator can "back up." A slight decrease in water pressure can result in a greater vacuum in your filtration apparatus, causing it to suck water back into the apparatus. If you see water entering the trap, break the vacuum by opening the stopcock or pinch clamp on the trap, and then turn off the water.

The actual filtration   For nonvolatile solvents like water, apply the vacuum and pour the crystallization mixture into the Büchner funnel at such a rate that the bottom of the funnel is always covered with some solution. For high-boiling organic solvents, pour an initial portion of the mixture into the funnel; then apply the vacuum. In both cases, when the vacuum is applied, the mother liquor is literally sucked through the filter paper into the filter flask while the crystals remain on the filter paper. When the mother liquor ceases to flow from the funnel stem, release the vacuum by opening the stopcock on the trap. Then turn off the aspirator or vacuum line.

Washing   To wash the contaminating mother liquor from the crystals, transfer the crystal mass, or filter cake, to a small beaker, using a spatula to loosen, remove, and scrape the filter paper. Place fresh filter paper in the Büchner funnel, stir the crystals with a small amount of *chilled* solvent, and then immediately refilter. Small amounts of crystals may be washed right in the funnel on the original filter paper. This procedure is not recommended because the wet filter paper may tear when you stir the wash solvent into the crystals and because this type of washing is not as thorough as a beaker washing.

Remove excess solvent from the crystals by putting a fresh piece of filter paper *on top of the crystals* still in the funnel and pressing this filter paper down firmly and all over with a cork. Keep the vacuum on during this pressing. When as much solvent has been pressed out of the filter cake as possible, leave the vacuum running for another minute or so. The air pulled through the filter cake will remove even more solvent. Then, open the trap clamp, turn off the vacuum, remove the Büchner funnel, and disconnect the filter flask assembly from the vacuum line. Using a spatula, pry the filter cake from the funnel for drying. The filter cake will often adhere to wet filter paper. Scrape the crystals from the paper only after it has dried.

Do not discard the mother liquor (in the filter flask), but place it in a corked Erlenmeyer flask until the completion of the experiment. The reason for saving the mother liquor is that it may still contain a substantial amount of the desired compound. Until you can determine a percent recovery of yield, you will not know if it is worthwhile to attempt to recover more material.

5) Drying the crystals.   The filter cake removed from the Büchner funnel still contains an appreciable amount of solvent. The crystals must be dried thoroughly before they can be weighted or before a melting point can be taken.

There are many methods of drying crystals. The simplest is *air-drying*, in which the crystals (with any lumps crushed) are spread out on a watch glass or large piece of filter paper and allowed to dry. Air-drying is some-times slow, especially if water or some other high-boiling solvent was used. Unless the crystals are partially covered, they can collect dust. Another watch glass or a beaker, propped on corks to allow air to get to the crystals, may be used as a cover. For a melting-point determination, a few crystals may be removed from the mass and allowed to air-dry on a separate, uncovered watch glass. If a compound is hygroscopic (attracts water from the air), it cannot be air-dried.

In some laboratories, particularly in large student laboratories, drying chemicals on watch glasses in an open room that has a static atmosphere is discouraged for health reasons. Before air-drying your product, check the policy in force in your laboratory.

A **desiccator** may be used for drying a water-crystallized or hygroscopic compound. A desiccant (drying agent) such as anhydrous calcium chloride is placed in the bottom of the desiccator; the porcelain shelf is inserted; then a

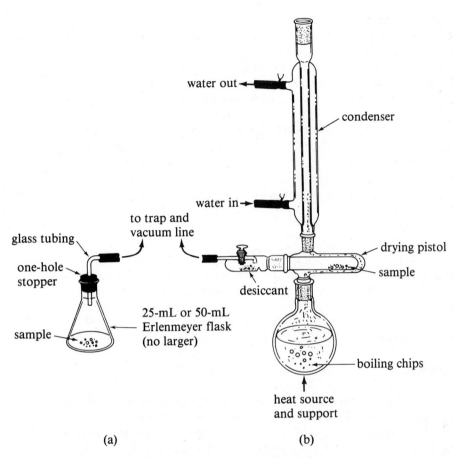

water out

condenser

water in

to trap and
vacuum line

glass tubing

one-hole
stopper

drying pistol

sample

desiccant

sample

25-mL or 50-mL
Erlenmeyer flask
(no larger)

boiling chips

heat source
and support

(a)   (b)

FIGURE 1.4 Two setups for drying small amounts of solid material under vacuum.
(a) A simple student setup. A corked sidearm test tube may be used similarly. (b) A
drying pistol. The glass tube holding the sample is heated by boiling water or other
solvent in the flask. In either type of setup, the Erlenmeyer or round-bottom flask
should be securely clamped.

watch glass or beaker holding the crystals is placed on the shelf. When the
cover is in place, the desiccant attracts water from the atmosphere in the
desiccator as the water evaporates from the crystals. Some desiccators can be
evacuated, thus speeding up the evaporation of any solvent. Figure 1.4
shows two setups for drying a small quantity of a solid under vacuum.

6) The second crop. The mother liquor from a crystallization may
still contain in solution a large amount of the desired compound. In many
cases, another batch of crystals, called the **second crop,** can be obtained from
this mother liquor. To get the second crop, boil away one-third to one-half of

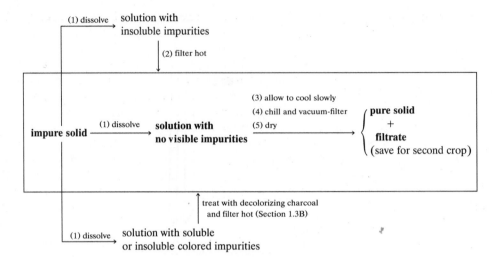

FIGURE 1.5 Steps in crystallization. (The numbers refer to the numbered steps in Section 1.2 of the text.)

the solvent from the mother liquor, and then allow it to cool and crystallize, as you did for the first crop of crystals. A seed crystal added to the cooled solution may help start the crystallization process. The second crop of crystals is rarely as pure as the first crop, because the impurities have been concentrated in the mother liquor. The purity of the second crop can be improved by recrystallizing these impure crystals with fresh solvent. Do not combine the first and second crops unless they are equal in purity.

The recovery in the second crop is usually reported separately from that in the first crop. In your notebook, you might write

% recovery (1st crop)   = 76%
% recovery (2nd crop) = 5.5%
total % recovery   = 81.5%

When no more crops are to be isolated, discard the crystallization solvent in a designated waste container. (See Appendix IV for a discussion of discarding organic material.) Figure 1.5 outlines the steps in crystallization.

1.3 Supplemental Procedures

A. Choosing a Solvent

If you do not know the appropriate solvent to use in crystallizing a compound, you must determine experimentally the solubility of the compound in various solvents. The following procedure can be used for this determination.

Weigh exactly 0.10 g of the compound into a small test tube. Pipet 1.0 mL of water into another test tube. Use these two test tubes as references for estimating 0.10 g of compound and 1.0 mL of solvent for the other test tubes.

Place an estimated 0.10 g of the compound into a series of small test tubes and add 1.0 mL of a different solvent to each. Try solvents of different polarities (Table 1.1). Stir each sample and determine the solubility of the compound in each solvent at room temperature. Record in your notebook whether the compound is insoluble (no apparent solid dissolved), slightly soluble (some of the solid dissolved), or soluble (no solid remains).

Place each test tube in a beaker of warm or boiling water, depending on the boiling point of the solvent. If some of the solvent should boil away, replace it with fresh solvent to maintain the volume at 1.0 mL.

Note the solubilities of the compound in the hot solvent. Some compounds may contain insoluble impurities even though the compound itself dissolves. Return the test tubes to a rack and allow them to cool to room temperature. Finally, place the test tubes in an ice bath. Record your observations.

From the preceding tests, choose the best solvent or solvent pair (Section 1.1); then weigh and crystallize the remainder of your sample, using about 10 mL of solvent per gram of solute. The compound in the test samples can be recovered by evaporating the test solvents in the fume hood and combining the solid material with the main sample before crystallization. (Evaporation of aqueous mixtures to recover material is usually not successful.)

B. Use of Decolorizing Charcoal

Frequently, small amounts of colored compounds and tarry (long-chain, or polymeric) materials are found as colored impurities in colorless organic compounds. These colored impurities can cause the crystallization solution and even the final crystals to have a tinge of color. These impurities can be removed with **decolorizing charcoal** (also called *activated charcoal* or *activated carbon*). The fine particles of carbon in decolorizing charcoal have a large surface area and *adsorb* organic compounds, especially colored and polymeric compounds. Add the decolorizing charcoal to the initial crystallization solution after the impure solid has been dissolved.

• •

SAFETY NOTE Never add decolorizing charcoal to a solution near its boiling point! The fine particles act like thousands of tiny boiling chips and will cause a hot solution to boil over.

• •

Only a small amount of decolorizing charcoal should be used; a "pinch" (0.1–0.5 g) on the end of a spatula is sufficient for most purposes. It is always better to err by adding too little charcoal than by adding too much, because the particles of carbon can adsorb the desired compound as well as impurities. However, if the activated carbon has lost much of its "activity" by exposure to air, considerably more will be needed to remove the color.

After adding the charcoal, swirl the mixture a few times, and then *carefully* reheat. (Boiling solutions containing decolorizing charcoal have a tendency to froth.) Filter the hot mixture through fluted, low-porosity filter paper, or filter it with a vacuum, using a filter aid (Section 1.3C).

Because decolorizing charcoal is messy and because its use always results in the loss of some of the compound being crystallized, decolorizing is carried out only when necessary and not as a routine procedure. If you are considering decolorizing charcoal, first test an aliquot of the solution and judge the result.

C. Use of Filter Aids

Gelatinous precipitates, such as those of metal hydroxides and metal oxides, are difficult to remove from solutions of organic compounds because they clog the filter paper. Finely divided contaminants, such as decolorizing charcoal, are also difficult to remove because they may pass through filter paper. To circumvent these problems, use a **filter aid** (Filter-Cel®, Celite®). Filter aids are useful only in removing a solid contaminant; they cannot be used for filtering of a desired solid.

Filter aids are diatomaceous earths (silica), which have large surface areas and are thus good adsorbents for contaminating solids. There are two techniques for using a filter aid: (1) Add the filter aid directly to the solution to be filtered, then heat or shake the mixture, and finally filter the mixture with vacuum. (2) Vacuum-filter a slurry of the filter aid and fresh crystallization solvent, and then filter the organic solution through the layer of filter aid resting on the filter paper. With either technique, use enough filter aid to cover the filter paper to a depth of 2–3 mm.

After filtering the solution, carefully wash the filter aid with fresh solvent to remove any adsorbed organic compound, refilter, and combine the filtrates.

Suggested Readings

Vogel, A. I. *Practical Organic Chemistry*. 3rd ed. London: Longman Group Limited, 1956.

Boschmann, E. "Efficient Use of Washing Solvents." *J. Chem. Ed.* **1972,** 49(9), 650.

Nilles, G. P., and Schuetz, R. D. "Selected Properties of Selected Solvents." *J. Chem. Ed.* **1973,** 50(4), 267. [See also Nilles, G. P., and Schuetz, R. D. *J. Chem. Ed.* **1973,** 50(12), 871.]

Baumann, J. B. "Solvent Selection for Recrystallization." *J. Chem. Ed.* **1979,** *56(1),* 64.

Weissberger, A., Proskauer, E. S., Riddick, J. A., and Toops, E. E. *Technique of Organic Chemistry VII. Organic Solvents.* 2nd ed. New York: Interscience Publishers, 1955. (Properties and purification of solvents.)

Problems

1.1 Each of the following compounds, A–C, is equally soluble in the three solvents listed. In each case, which solvent would you choose? Give reasons for your answer. (More than one answer may be correct.)
(a) Compound A: benzene, acetone, or chloroform
(b) Compound B: carbon tetrachloride, dichloromethane, or ethyl acetate
(c) Compound C: methanol, ethanol, or water

1.2 Which of the following solvents could *not* be used as solvent pairs for crystallization? Explain.
(a) Ligroin and water
(b) Chloroform and diethyl ether
(c) Acetone and methanol

1.3 Suggest possible crystallization solvents for the following compounds.

(a) Naphthalene, (mp 80°)

(b) Succinic acid, $HOC(CH_2)_2COH$ (mp 188°)

(c) *p*-Iodophenol, I—⟨ ⟩—OH (mp 94°)

1.4 Give reasons for each of the following experimental techniques used in crystallization.
(a) A hot crystallization solution is not filtered unless absolutely necessary.
(b) An Erlenmeyer flask containing a hot solution is not tightly stoppered to prevent solvent loss during cooling.
(c) The suction of a vacuum filtration apparatus is broken before the vacuum is turned off.
(d) Vacuum filtration is avoided when crystals are isolated from a very volatile solvent.
(e) Carborundum (silicon carbide) boiling chips are better than calcium carbonate chips in the crystallization of an unknown.

1.5 A chemist crystallizes 17.5 g of a solid and isolates 10.2 g as the first crop and 3.2 g as the second crop.
(a) What is the percent recovery in the first crop?
(b) What is the total percent recovery?

1.6 A student crystallized a compound from benzene and observed only a few crystals when the solution cooled to room temperature. To increase the yield of crystals, the student chilled the mixture in an ice–water bath. The chilling greatly increased the quantity of solid material in the flask. Yet when the student filtered these crystals with vacuum, only a few crystals remained on the filter paper. Explain this student's observations.

1.7 How could you determine the minimum amount of decolorizing charcoal needed to decolorize a crystallization solution?

1.8 The solubility of acetanilide in hot water (5.5 g/100 mL at 100°) is not very great, and its solubility in cold water (0.53 g/100 mL at 0°) is significant. What would be the maximum theoretical percent recovery (first crop only) from the crystallization of 5.0 g of acetanilide from 100 mL of water (assuming the solution is chilled to 0°)?

Melting Points

The **melting point** of a crystalline solid is the temperature at which the solid changes to a liquid at 1.0 atmosphere of pressure. The melting point is the same as the freezing point, the temperature at which the liquid becomes solid. Because liquids have a tendency to become supercooled (remain liquid below their freezing points), freezing-point determinations are only rarely performed in organic chemistry.

2.1 Characteristics of Melting Points

The melting point of a solid should be reported as a **melting range.** The barometric pressure, which has a negligible effect on the melting points at the usual atmospheric pressures, is ignored.

The melting point is determined by heating a small sample of the solid material slowly (at the rate of about 1° per minute). The temperature at which the first droplet of liquid is observed in the solid sample is the lower temperature of the melting range. The temperature at which the sample finally becomes a clear liquid throughout is the upper temperature of the melting range. Thus, a melting point might be reported, for example, as mp 103.5°–105°.

A. Effect of Impurities

A pure organic compound usually has a "sharp" melting point, which means that it melts within a range of 1.0° or less. A less pure compound exhibits a broader range, maybe 3° or even 10°–20°. For this reason, a melting point can often be used as a criterion of purity. A melting range of 2° or less indicates a compound pure enough for most purposes. However, a compound purified for spectroscopic analysis or for submission to an analytical laboratory for elemental analysis (determination of the relative weight percentages of the elements) should have a very sharp melting point.

An impure organic compound exhibits not only a broad melting range but also a *depressed* (lower) melting point from that of the pure compound. For example, a fairly pure sample of benzoic acid might melt at 121°–122°, but an impure sample might show a melting range of 115°–119°.

Using Melting Points to Identify Unknowns

The melting point of a compound can be used to prove the identity of a pair of compounds. Assume that you have a compound of unknown structure that melts at 120°–121°. Is the compound benzoic acid? To find the answer, you would mix the unknown with an authentic sample of benzoic acid (mp 120°–121°), and then take a melting point of the mixture. This melting point is called a **mixed melting point**. If the unknown is benzoic acid, the mixed melting point would remain 120°–121°, because the two mixed samples are the same compound. However, if the unknown is *not* benzoic acid, the mixed melting point would be depressed and show a wider range. For absolute identification purposes, additional data besides the mixed melting point are desirable.

Just comparing the melting point of an unknown with a literature value is insufficient evidence to identify an unknown. There are literally hundreds of different compounds that have the same melting point. However, even if two different compounds have the same melting point, their mixed melting point will be broad and depressed.

B. Melting-Point Diagrams

The melting-point diagram in Figure 2.1 shows the typical melting behavior of a series of mixtures of two organic compounds, *o*-dinitrobenzene and *m*-dinitrobenzene.

o-dinitrobenzene
(1,2- or *ortho*-dinitrobenzene)

m-dinitrobenzene
(1,3- or *meta*-dinitrobenzene)

A similar graph for mixtures of two other compounds would probably be similar, although different types of melting behavior are occasionally observed (Section 2.1C). The graph shows only the upper limit of the melting range, the temperature at which the mixture becomes completely liquid. The graph also shows that pure *o*-dinitrobenzene melts at 118.5°, pure *m*-dinitro-

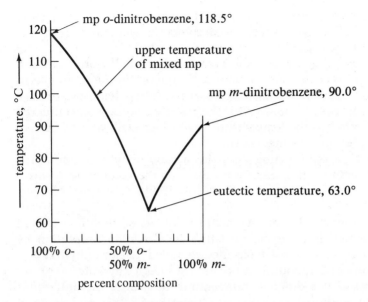

FIGURE 2.1   Melting-point diagram for mixtures of o-dinitro-benzene and m-dinitrobenzene.

benzene melts at 90.0°, and a 50:50 (molar ratio) mixture of the two compounds melts at about 80°.

The lower point on the graph (63°) is called the **eutectic point,** or **eutectic temperature,** and it is the lowest melting point of any mixture of these two compounds. The percent composition of the mixture that melts at the eutectic point is called the **eutectic mixture.** The values for the eutectic point and percent composition of a eutectic mixture depend on which compounds are being studied. For the diagram in Figure 2.1, the eutectic mixture consists of 63% m-dinitrobenzene and 37% o-dinitrobenzene. Binary mixtures of other compounds will have different values for their eutectic temperature and mixture. The eutectic point for any binary mixture is the temperature at which both components melt simultaneously and is, consequently, a sharp melting point rather than the broad melting range usually observed for mixtures.

A eutectic point is exhibited only by an intimate and uniform mixture of the correct composition. For all practical purposes in the organic laboratory, virtually all mixtures of two different compounds exhibit a broad melting range.

C. Other Melting Behavior

Decomposition   All organic compounds decompose when heated to sufficiently high temperatures. Many organic compounds decompose at or below their melting points. Some of these compounds exhibit sharp melting points with evidence of decomposition, such as darkening. Others, even pure compounds, may exhibit a broad melting–decomposition range.

Polymorphs   Some compounds exhibit polymorphism. Polymorphs are different crystalline forms of the same substance. Variations in the intermolecular attractions of the different crystalline forms can cause polymorphs to have different, but sharp, melting points. Which crystalline form is encountered depends on such factors as the temperature at which crystallization occurred, the rate of crystallization, and the solvent.

When more than one melting point for a pure organic compound is reported in the literature, it generally means that the compound forms a polymorph.

Hydrates   Some compounds can crystallize with water or other solvent molecules incorporated in their crystal lattice in a definite proportion by weight. In the case of water, these molecules are called "water of hydration," and the combination of compound and water is called a "hydrate."

A hydrate melts at a different temperature than does the anhydrous form of the compound. For example, the hydrate of oxalic acid melts at 101.5°, but anhydrous oxalic acid exists in two polymorphic forms, one melting at 182° and the other at 189.5°.

Racemic mixtures and racemates   A pair of nonsuperposable mirror image isomers are called **enantiomers.** (Consult your lecture text for more details.)

$$R_4 - \overset{\overset{\displaystyle R_1}{|}}{C} - R_2 \qquad R_2 - \overset{\overset{\displaystyle R_1}{|}}{C} - R_4$$

$$\underset{R_3}{} \qquad \underset{R_3}{}$$

$$\text{A} \qquad\qquad \text{B}$$

Let us call a hypothetical pair of enantiomers A and B. Because of structural similarities, pure A and B have identical melting points.

When a 50 : 50 mixture of enantiomers is crystallized, each enantiomer may crystallize separately. In our example, we would have a 50 : 50 mixture of A crystals and B crystals, which is called a **racemic mixture.** A racemic mixture does not melt at the same temperature as pure A or pure B. Physically, this mixture is like any mixture of different compounds and exhibits a depressed melting point.

In some cases, however, a pair of enantiomers crystallize so that each crystal contains both A and B in a regular alternating pattern. This type of mixture is called a **racemic compound,** or **racemate.** A racemate behaves as a pure compound, with its own sharp melting point. The melting point of a racemate may be higher or lower than that of either pure enantiomer (Figure 2.2). The melting point differs from that of either pure enantiomer because the crystalline structure is different.

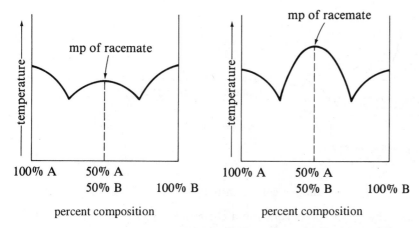

FIGURE 2.2   Melting-point diagrams for two racemates, each showing two eutectic points. In the first diagram, the racemate is lower melting than pure A or B. In the second diagram, the racemate is higher melting than either A or B.

Different crystalline structures explain why more than one melting point is often reported for compounds of biological interest. For example, the *Handbook of Chemistry and Physics* lists the melting point of *d*-menthol as 42°–43° and the melting point of *dl*-menthol as 28°.

Liquid crystals   Most pure organic compounds melt over a 1°–2° temperature range. A wide melting range usually arises from impurities, but this is not always the case. About 0.5% of pure organic compounds do not melt sharply but form a **mesophase** before melting (Greek *meso*, "middle" or "intermediate"). The mesophase, which is intermediate between the solid phase and the liquid phase, is called a **liquid crystal.** Liquid crystals that are formed by heat are called **thermotropic liquid crystals.** The term *liquid crystal*, unfortunately, is now used to refer to both the compound that melts to form a mesophase and to the mesophase itself.

In a melting-point apparatus, a compound forming a liquid crystal phase does not show a sharp melting point. Instead, we observe two ill-defined transitions, each at its own characteristic temperature. In most cases, a transition from the solid to a turbid liquid is observed at one temperature, and at a higher temperature, another transition to a clear liquid is seen.

$$\underset{\substack{\text{solid} \\ \text{(a liquid crystal)}}}{} \xrightarrow{\substack{\text{first transition} \\ \text{temperature}}} \underset{\substack{\text{mesophase} \\ \text{(the liquid crystal phase)}}}{} \xrightarrow{\substack{\text{second transition} \\ \text{temperature}}} \text{liquid}$$

In the liquid crystal mesophase, the molecules are more orderly than those in a liquid but less orderly than those in a solid. The types of molecules

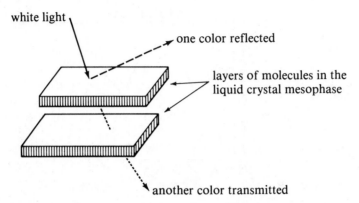

white light

one color reflected

layers of molecules in the
liquid crystal mesophase

another color transmitted

FIGURE 2.3   The interaction of light with a cholesteric liquid
crystal. The helical structure of the liquid crystal is not
depicted here.

that can form liquid crystals are generally large organic molecules that are
flat and elongated, somewhat like flattened rods. Many of these compounds
are polar. In the liquid crystal mesophase, these rodlike molecules are
aligned parallel to one another, much like a bundle of pencils, but can move
around.

   Cholesteric liquid crystals are a type of liquid crystal in which the parallel
molecules are in layers slightly displaced from one another. The displacement of the layers arises from side chains that protrude from the otherwise
flat molecules. Because of the displacement, the layers form a helix resem-
bling a spiral staircase. (The term *cholesteric* is derived from the chemical
name "cholesterol," because this type of liquid crystal is commonly formed
by derivatives of cholesterol.)

   The unusual arrangement of liquid crystal molecules results in some
very interesting optical properties, some of which find application in electronic devices such as watch displays and digital thermometers.* When
white light, which contains all visible wavelengths, strikes the surface of a
cholesteric liquid crystal, the light is separated into two components. Both
components are colored because they arise from only part of the original
white light. Light of one color is reflected from the surface, while light of
another color is transmitted through the liquid crystal (Figure 2.3). The result
is a vivid display of iridescence.

   A small change in temperature results in a small change in the pitch of
the helix (distance required for one complete revolution of the helix). This
change in pitch alters the colors of light that are reflected and transmitted, a
property that is useful in liquid crystal thermometers and novelty items.

---

*For an article concerning the structures and uses of liquid crystals, see G. H. Brown and P. P.
Crooker, *Chemical and Engineering News*, Jan. 31, 1983, page 24.

2.2 Melting-Point Apparatus

Many types of electrically heated melting-point devices are on the market today; two are shown in Figure 2.4. With one type of device, a glass capillary tube containing the sample is inserted into a heating block, and the melting of the compound is observed through a magnifying eyepiece. Most instruments will accept several capillary tubes simultaneously, so that the melting points of several samples can be determined under identical conditions. Simultaneous melting-point determinations are especially useful for mixed melting points, because the melting behavior of the pure compound and of one or more mixtures can be compared.

A *melting-point stage* is an apparatus in which the sample (as little as a single crystal) is placed between two microscope cover glasses instead of in a

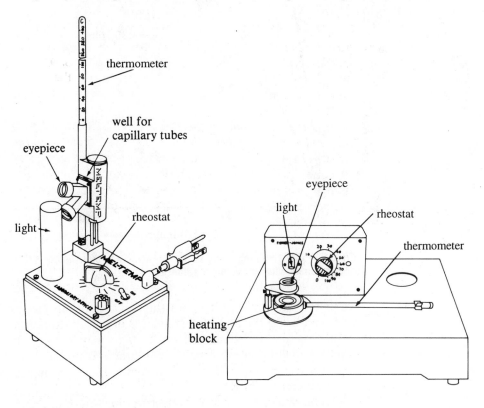

capillary melting-point device   melting-point stage

FIGURE 2.4 Two types of melting-point devices. Melting-point capillaries containing samples are inserted into the capillary well in the left-hand device. Crystals are placed between microscope cover glasses and put on the heating block in the right-hand device.

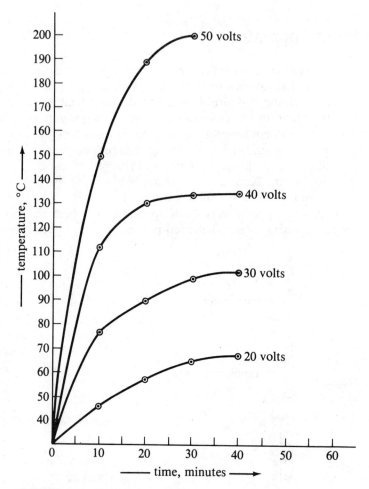

FIGURE 2.5 Heating curves for a Mel-Temp® melting-point apparatus at different voltage settings.

capillary. The cover glasses are then placed on an electrically heated block (the stage), and the melting behavior is observed through a magnifying glass. A melting point determined on a stage may be slightly higher than the capillary melting point, because the sample is not actually submerged in the heat source. In the following instructions, we assume you will be taking capillary melting points.

The rate of heating of an electrical melting-point apparatus is controlled by a rheostat. The higher the setting on the rheostat dial, the faster the block will heat. The rate of heating at different dial settings can be measured and graphed. Figure 2.5 shows typical heating curves for one type of commercial instrument. Note that the rate of heating at any one voltage setting is not linear but drops off with increasing temperature. This variable rate is advantageous because at one voltage setting, the sample can be heated rapidly at first and then more slowly as the melting point is approached.

2.3 Steps in Determining a Melting Point

1) Preparation of the sample. Pulverize the sample for a melting-point determination by placing 0.1–0.2 g of *dry* crystals in a watch glass and crushing them with a metal spatula or the end of a test tube. If the sample is to be used for a mixed melting point, grind a 50:50 mixture of the two compounds (approximated, not necessarily weighed) *thoroughly* with a mortar and pestle to ensure a homogeneous and intimate mixture. Alternatively, dissolve the mixture of the two compounds in a suitable solvent on a watch glass; allow the solvent to evaporate; and then pulverize the residue.

2) Loading the capillary. Mound up the pulverized sample and press the open end of the capillary into the sample against the surface of the watch glass or mortar. A small plug of sample will be pushed into the open end of the capillary tube. The ideal amount of sample is a plug about 1 mm in length. Tapping the sealed end of the capillary on the surface of the laboratory bench may knock the sample down to the desired position at the sealed end of the capillary. (CAUTION: Capillaries are fragile.) A safer and more efficient technique for driving the plug of sample to the sealed end of the capillary tube is to drop the tube (sealed end down) through 2–3 feet of ordinary glass tubing vertically onto the benchtop. The impact of the capillary with the hard surface knocks the sample down. It may be necessary to drop the capillary two or three times.

It is important that the height of the sample in the capillary be only 1–2 mm and that it be packed firmly. A larger sample takes longer to melt and may exhibit an erroneously large melting range. If the sample is loosely packed, it is difficult to determine the start of the melt.

3) Preliminary melting point. If the approximate melting point of a sample is not known, it saves time to take a preliminary melting point with a second capillary tube. The approximate melting point of the sample is determined by rapidly heating the capillary with the melting-point apparatus (about 10° per minute). This preliminary melting point will be on the low side of the true melting point, but it will give you an idea of the rate of temperature increase you should use to determine the exact melting point.

You need not determine a preliminary melting point if you know the name of the compound and can find its melting point in a reference book or journal. For example, the *Handbook of Chemistry and Physics* (CRC Press, Inc., Boca Raton, Florida) contains an extensive section called "Physical Constants of Organic Compounds," along with instructions for finding compounds in the listing. Another good reference for physical constants is *Heilbron's Dictionary of Organic Compounds* (Oxford University Press, New York).

4) *Taking the melting point.* Insert the capillary tube containing the sample into the melting-point apparatus, along with the thermometer, if necessary. Heat the apparatus rapidly to a temperature about 10° below the expected melting point, then slow the rate of temperature increase to about 1° per minute. If the temperature is increased too rapidly at the melting point, the thermometer, sample, and block will not be at thermal equilibrium, and erroneous readings will result.

As we have mentioned, the initial, or lower, end of the melting range is the temperature at which the first drop of liquid is noted in the solid sample. The final, or upper, end of the range is the temperature at which the sample becomes a clear liquid containing no solid material. The determination of the final value of the melting range usually presents no problems. The initial value, however, may require judgment. Many organic compounds undergo changes in crystal structure prior to melting, and these phase changes may be mistaken for the first sign of melting. Sample sag, shrinking, changes in texture, and the appearance of droplets *outside the bulk of the sample* are not the start of the melt. The initial temperature of the melting range is taken as the first appearance of liquid *within the bulk of the sample*.

Another problem that may arise is *decomposition*. In a simple case, a sample may change color, effervesce, or otherwise change in appearance at its melting point. If the melting point is reasonably sharp, this type of behavior does not affect the value of the melting point as a criterion of purity or as an identification tool, but the decomposition should be reported along with the melting point—for example, mp 150.3°–151.5°d; 150.3°–151.5° dec; or 150.3°–151.5° with darkening. If a sample decomposes over a large temperature range, the melting point cannot be used for identification purposes. The decomposition range should be reported, even if only approximate temperatures can be given. For example: dec 127°– ~150°.

After taking a melting point, discard the used, cooled capillary tube; the tube cannot be cleaned for reuse.

2.4   Supplemental Procedures

A. Melting Point of a Volatile Solid

Some solids (such as camphor or menthol) vaporize before melting when they are heated. To determine the melting point of a volatile solid, the open end of the melting-point capillary must be sealed before the melting point can be taken.

To seal a melting-point capillary, first load the capillary in the usual way. Then, holding the capillary tube horizontally, twirl or roll the capillary

TABLE 2.1 Suggested melting-point standards
for calibration of thermometers

Compound	Melting point (°C)
naphthalene	80.6
1,3-dinitrobenzene	90.0
acetanilide	114.3
benzoic acid	122.4
benzamide	130
urea	135
sulfanilamide	166
p-toluic acid	182
succinic acid	188
3,5-dinitrobenzoic acid	205
anthracene	216.3
p-nitrobenzoic acid	242

with your fingers, having about a millimeter of the open end in the base of a flame. The edges of the open end will melt and collapse. The twirling or rolling of the tube back and forth ensures that the glass will collapse inward and seal the tube. After sealing the tube, inspect the closed end with a magnifying glass to be sure that it does not contain a pinhole.

B. Thermometer Calibration

No thermometer is accurate at every temperature reading; therefore, a thermometer to be used for melting-point determinations in a research laboratory (and sometimes in the student laboratory) should be calibrated. Calibration is accomplished by recording the melting points of five or six very pure compounds, chosen to melt at a variety of temperatures. Table 2.1

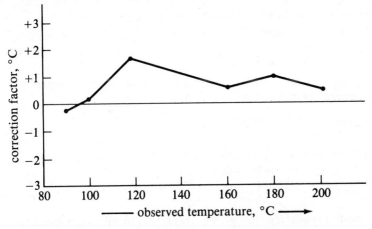

FIGURE 2.6 A typical thermometer calibration graph.

lists some suggested compounds. From these melting points, a graph similar to the one in Figure 2.6 is constructed. This graph shows the *correction factor* versus the *observed temperature* (using the upper value of the melting range). For example, pure benzoic acid melts at 122.4°. If your thermometer records 120.9°, your correction factor at approximately 120° would be +1.5°. Any time you record a melting point near 120°, you would add 1.5° to the observed temperature. (Corrected melting point = observed melting point + 1.5°.) A corrected melting point is reported as follows: mp 121.2°–121.5° (cor).

Melting points taken with a thermometer calibrated as described need no additional correction factor for the exposed thermometer stem. If you use a thermometer that has been factory-calibrated while completely submerged, you may want to correct the melting points with *stem correction factors*. Calculations of these correction factors are discussed in the texts by Morton and by Vogel listed in the Suggested Readings at the end of this section.

Suggested Readings

Morton, A. A. *Laboratory Techniques in Organic Chemistry*. New York: McGraw-Hill, 1938, Chapter 2.

Vogel, A. I. *Practical Organic Chemistry*. 3rd ed. London: Longman Group Limited, 1956.

Weissberger, A., ed. *Technique of Organic Chemistry I. Physical Methods*. 3rd ed. New York: Interscience Publishers, 1959, Chapter 7.

Allen, E. "The Melting Point of Impure Organic Compounds." *J. Chem. Ed.* **1942,** *19(6)*, 278.

Problems

2.1 Using the heating curves in Figure 2.5, what should be your rheostat settings for a compound expected to melt at (a) 150°; (b) 110°?

2.2 Using the *Handbook of Chemistry and Physics,* find the expected melting points of the following compounds.
(a) Indophenol   (b) Quinine

2.3 (a) Using the following table, draw a melting-point diagram and estimate the eutectic temperature and composition.

Percent composition	Melting range
100% A	125°–126°
75% A–25% B	115°–120°
65% A–35% B	128°–131°
50% A–50% B	135°–140°
100% B	151°–152°

(b) If you wished to obtain a more accurate graph for this melting-point curve, suggest the best mixtures for additional mixed melting points.

2.4 A compound is observed to melt sharply at 111° with the vigorous evolution of a gas. Then the compound solidifies and does not melt until 155°, at which temperature it again melts sharply. Explain.

2.5 Suppose that your sample melts before you are ready to record the melting point. Should you (a) cool the capillary and redetermine the melting point, or (b) begin with a fresh sample? Explain.

2.6 Why is it *not* good practice to mix two components by simply stirring them together and then sampling them for a mixed melting point?

2.7 What conclusions would you draw from the following student observations? The melting point of unknown sample A is 83.0°–89.5°. The melting point of an authentic sample of 1-naphthol is 93.5°–94.0°. A mixed melting point of these two samples is observed to be 88.0°–91.5°.

2.8 Using Table 2.1, construct a thermometer calibration graph from the following observed melting points.
Acetanilide, mp 113°
Benzamide, mp 130°
Succinic acid, mp 188°
p-Nitrobenzoic acid, mp 245°

Extraction Using
a Separatory Funnel

Extraction is the separation of a substance from one phase by another phase. The term is usually used to describe removal of a desired compound from a solid or liquid mixture by a solvent. In a coffee pot, caffeine and other compounds are extracted from the ground coffee beans by hot water. Vanilla extract is made by extracting the compound vanillin from vanilla beans.

In the laboratory, several types of extraction techniques have been developed. The most common of these is *liquid–liquid extraction,* or simply "extraction." Extraction is often used as one of the steps in isolating a product of an organic reaction. After an organic reaction has been carried out, the reaction mixture usually consists of the reaction solvent and inorganic compounds, as well as organic products and by-products. In most cases, water is added to the reaction mixture to dissolve the inorganic compound. The organic compounds are then separated from the aqueous mixture by extraction with an organic solvent that is immiscible with water. The organic compounds dissolve in the extraction solvent, while the inorganic impurities remain dissolved in the water.

The most commonly used device to separate the two immiscible solutions in an extraction procedure is the **separatory funnel.** Typically, the aqueous mixture to be extracted is poured into the funnel first, and then the appropriate extraction solvent is added. The mixture is shaken (with the precautions discussed in Section 3.3) to mix the extraction solvent and the aqueous mixture, and then it is set aside for a minute or two until the aqueous and organic layers have separated. The stopcock at the bottom of the separatory funnel allows the bottom layer to be drained into a flask and allows the separation of the two layers, as shown in Figure 3.1. The result (ideally) is two separate solutions: an organic solution (organic compounds dissolved in the organic extraction solvent), and an inorganic solution (inorganic compounds dissolved in water). Unfortunately, often the water layer still contains some dissolved organic material. For this reason, the

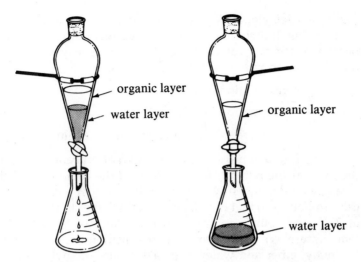

FIGURE 3.1 Two immiscible solutions can be separated with a separatory funnel. (The organic layer may be the upper or lower layer, depending on the relative densities of the two solutions.)

water layer is usually extracted one or two times with fresh solvent to remove more of the organic compound. In addition, some of the water layer is usually retained on the sides of the separatory funnel and contaminates the organic layer. This water contaminant is removed in a separate step later in the workup or isolation procedure (see Technique 4).

After one or more extractions and separations, the combined organic solutions are usually extracted with small amounts of fresh water to remove traces of inorganic acids, bases, or salts; treated with a solid drying agent to remove traces of water; and then filtered to remove the hydrated drying agent. Finally, the solvent is evaporated or distilled. The organic product can then be purified by a technique such as crystallization or distillation.

3.1 Distribution Coefficients

When a compound is shaken in a separatory funnel with two immiscible solvents, such as water and diethyl ether ($CH_3CH_2OCH_2CH_3$), the compound distributes itself between the two solvents. Some of compound dissolves in the water and some in the ether. How much of the compound or solute dissolves in each phase depends on the solubility of the solute in each solvent. The ratio of the concentrations of the solute in each solvent at a particular temperature is a constant called the **distribution coefficient** or

partition coefficient (K). (In calculations of distribution coefficients, we assume that the solute neither ionizes in nor reacts with either solvent.) Because a ratio is involved, the concentrations may be in any units, as long as the two concentrations are the *same* units.

$$K = \frac{\text{concentration in solvent}_2}{\text{concentration in solvent}_1}$$

where solvent$_1$ and solvent$_2$ are immiscible liquids

To a rough approximation, the ratio of concentrations of this equation is the same as the ratio of the *solubilities* of the compound in the two solvents, measured independently. For example, consider a compound, which we will call compound A, that is soluble in diethyl ether to the extent of 20 g/100 mL at 20°C and soluble in water to the extent of 5.0 g/100 mL at the same temperature. We can approximate the distribution coefficient of compound A in diethyl ether and water to be 4.0, where diethyl ether is solvent$_2$ in the following equation:

$$K = \frac{\text{solubility in solvent}_2}{\text{solubility in solvent}_1}$$

$$= \frac{20 \text{ g}/100 \text{ mL}}{5.0 \text{ g}/100 \text{ mL}}$$

$$= 4.0$$

Given the distribution coefficient for a particular system, we can calculate how much compound will be extracted. Supposed that we have a solution containing 5.0 g of compound A in 100 mL of water. If we shake this solution with 100 mL of diethyl ether, how much A will be extracted by the ether? For this calculation, we will assume that the ether–water distribution coefficient is 4.0.

To solve this problem, we will let x be equal to the number of grams of compound A in the diethyl ether. Since we started with 5.0 g of A in the water, the number of grams remaining in the water is $5.0 - x$. The respective concentrations are the number of grams in each layer divided by each volume.

$$\text{concentration of A in diethyl ether} = \frac{x \text{ g}}{100 \text{ mL}}$$

$$\text{concentration of A in water} = \frac{(5.0 - x)\text{g}}{100 \text{ mL}}$$

Substitution,

$$K = \frac{\text{concentration in diethyl ether}}{\text{concentration in water}}$$

$$4.0 = \frac{x \text{ g}/100 \text{ mL}}{(5.0 - x)\text{g}/100 \text{ mL}}$$

Solving,

$$4.0 = \frac{x}{5.0 - x}$$

$$20 - 4.0x = x$$

$$5.0x = 20$$

$$x = 4.0 \text{ g in 100 mL of diethyl ether}$$

From this calculation, we see that we will extract only 80% of A from the water solution. Therefore, 20%, or 1.0 g, of A will remain in the water layer.

Because a solute distributes itself between two solvents, a single extraction may not be very efficient. Considerable amounts of material may remain behind in the original solvent. Is it possible to make liquid–liquid extraction more efficient? Using the same amount of diethyl ether, can we extract more than 4.0 g of A from the original water solution? Yes, we can. Let us divide the 100 mL of ether into three portions of approximately 33 mL each. Then, let us extract the water solution three separate times, using 33 mL of fresh diethyl ether for each extraction. If you carry out the calculation for each extraction as before (allowing for the different volumes of solvent), you will find that a total of 4.5 g of A can be extracted.

Our conclusion is that it is a more efficient use of solvent to perform three small extractions than one large one. A greater number of small extractions would remove an even greater quantity of the solute from water. If compound A were valuable, we would perform the extra extractions; otherwise, we would not bother.

Most organic compounds have distribution coefficients between organic solvents and water greater than 4. Therefore, a double or triple extraction generally removes almost all of the organic compound from the water. However, for water-soluble compounds, where K may be less than 1, we can calculate that only a small amount of the compound will be extracted. We predict that the extraction will fail. In order to extract the compound, we will have to use continuous liquid–liquid extraction (Section 3.6A) or salting out (Section 3.4B).

Immiscible

· · · · · · · · · · ·
3.2 Extraction Solvents

The preceding discussion has provided some clues for the choice of an extraction solvent. The extraction solvent must be immiscible with the first solvent, which is generally water. The compound to be extracted should be soluble in the extraction solvent and, of course, not undergo reaction with it. (Section 1.1 contains a brief discussion of solubilities.) Major impurities should not be soluble in the extraction solvent. In addition, the extraction solvent should be sufficiently volatile that it can be removed by distillation from the extracted material later in the workup procedure.

It is also preferable that the solvent be nontoxic and nonflammable. Unfortunately, these last two criteria are not met by many organic solvents.

TABLE 3.1   Some common extraction solvents

Name	Formula	Density (g / mL)[a]	bp (°C)	Comments
lighter than water:				
diethyl ether	$(CH_3CH_2)_2O$	0.7	35	highly flammable
petroleum ether	—[b]	~0.7	30–60	flammable
ligroin	—[b]	~0.7	>60[b]	flammable
benzene	C_6H_6	0.9	80	flammable, toxic, carcinogenic
toluene	$C_6H_5CH_3$	0.9	111	flammable
heavier than water:				
methylene chloride (dichloromethane)	CH_2Cl_2	1.3	41	toxic
chloroform (trichloromethane)	$CHCl_3$	1.5	61	toxic
carbon tetrachloride (tetrachloromethane)	CCl_4	1.6	77	toxic, carcinogenic

[a]The density of water is 1.0 g/mL; that of saturated aqueous sodium chloride solution is 1.2 g/mL.
[b]See Table 1.1, page 40.

Diethyl ether, the most common extraction solvent, is both volatile (boiling point only a few degrees above room temperature) and flammable. Benzene is toxic and flammable. Halogenated hydrocarbons are not all flammable, but most are toxic. When using a solvent for extraction, always proceed with caution. Table 3.1 lists a few common extraction solvents, their densities, and their potential hazards.

Note in Table 3.1 that the chlorinated hydrocarbon solvents are more dense than water; these solvents sink to the bottom in a separatory funnel containing water. The other solvents listed usually float on water. An exception would be a solvent containing a high concentration of a dense solute, which can increase the density of the organic layer so that it becomes heavier than water. Prematurely discarding the wrong solution is a common error. For this reason, it is generally wise to test the two layers if there is any question as to which is the organic layer and which is the aqueous layer. A simple test is to add a few drops of each layer to a small amount of water in a pair of test tubes. The layer that is immiscible with water is the organic layer. Also, it is generally wise to save all layers until an experiment is complete.

3.3   Steps in Extraction

1) Preparing the separatory funnel.   Lightly grease the glass stopcock of the separatory funnel with stopcock grease. (A Telfon® stopcock should not be greased.) Use just enough grease to glaze the clean ground

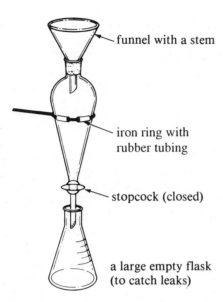

FIGURE 3.2 The appearance of the separatory funnel before liquids are poured into it.

glass when the joint is rotated. Too much grease will contaminate the organic solution and may even clog the stopcock. (If grease does get into the stopcock hole, remove it with a pipe cleaner or a rolled piece of paper.) Figure 3.2 shows the separatory funnel as it is positioned in an iron ring just prior to the addition of a liquid.

SAFETY NOTE Be sure the stopcock and funnel are matched. In student laboratories, the separatory funnels and their stopcocks are commonly separated for storage in the storeroom. If the stopcock and funnel are mismatched, no matter how well you grease the stopcock, it will leak. A simple check can be made by trying to align the borehole in the stopcock and the stem of the funnel. If they will not align, they are mismatched.

2) Adding the liquids. Be sure the stopcock is closed. Using an ordinary long-stem funnel, pour the solution to be extracted into the separatory funnel, followed by a measured amount of extraction solvent. Do not fill the separatory funnel more than three-quarters full; otherwise, there will not be enough room for mixing the liquids.

Never add a volatile solvent to a warm solution. Use caution when mixing any materials that can evolve a gas. If you are using a flammable solvent, make sure there are no flames in the vicinity!

3) Mixing the layers. Before inserting the stopper, swirl the separatory funnel gently. The swirling is especially important if an acid and sodium bicarbonate are present, because gaseous carbon dioxide is given off by their reaction. The swirling will drive off some of the carbon dioxide and minimize pressure buildup during the shaking process.

Insert the stopper and, *holding the stopper in place with one hand*, pick up the separatory funnel and invert it. Immediately open the stopcock with your other hand to vent solvent fumes or carbon dioxide. Swirl the inverted separatory funnel gently with its stopcock open to further drive off solvent vapors or gases.

Always aim the stem of the separatory funnel away from your neighbors when venting. Better yet, aim the stem into the hood.

Figure 3.3 shows the proper method for holding a separatory funnel. Hold the stopper and stopcock firmly in place throughout the entire shaking process to prevent their falling out.

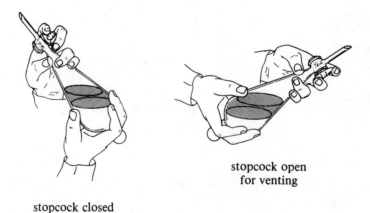

stopcock closed
for shaking

stopcock open
for venting

FIGURE 3.3 How to hold and vent a separatory funnel.

After venting, close the stopcock, gently shake or swirl the mixture in the inverted funnel, and then revent the fumes. If excessive pressure buildup is not observed, shake the separatory funnel and its contents up and down vigorously in a somewhat circular motion for 2–3 minutes so that the layers are thoroughly mixed. Vent the stopcock several times during the shaking period. After completing the shaking, vent the stopcock one last time. With the stopcock closed, place the separatory funnel back in the iron ring and remove the stopper. If you are extracting a small amount of material, wash the stopper into the separatory funnel with a few drops of extraction solvent, using a dropper. Place a large Erlenmeyer flask under the stem of the separatory funnel in case the stopcock should develop a leak. Allow the separatory funnel to sit until the layers have separated.

4) _Separating the layers._   Before proceeding, make sure the stopper has been removed. (It is difficult to drain the lower layer from a stoppered funnel, because a vacuum is created in the top portion of the funnel.) Partially open the stopcock to drain the lower layer into the flask. Hold the stopcock in place and brace the separatory funnel with both hands so that the stopcock cannot accidentally slip out of place (see Figure 3.4). Splashing is minimized if the tip of the stem touches the side of the flask so that the liquid can run down its side.

When the lower layer is almost, but not quite, drained into the flask, close the stopcock and gently swirl the separatory funnel. The swirling knocks drops clinging to the sides of the funnel to the bottom, where they can be removed. Carefully and slowly drain the last of the lower layer into

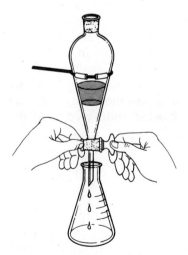

FIGURE 3.4 How to hold a separatory funnel while draining the lower layer.

the flask. Finally, tap the stem of the funnel to knock any clinging drops into the flask.

At this point in a synthesis experiment, the organic layer containing the desired product would be washed and poured from the top of the funnel into a clean Erlenmeyer flask. In this way, the upper layer is not contaminated with drops remaining in the stem.

5) Cleaning the separatory funnel. Clean the separatory funnel as soon as you are finished so that the organic solvent will not dissolve the stopcock grease and cause the stopper or stopcock to freeze. Then regrease glass stopcock or store it separately. A Teflon® stopcock should be washed but not greased and replaced *loosely* in the separatory funnel.

When the experiment is completed, the extracted aqueous layer and the organic layer should be discarded. Check with your instructor for the procedure to use to discard chemicals in your laboratory.

3.4 Additional Techniques Used in Extractions

A. Multiple Extractions

In practice, a single extraction is rarely used; multiple extractions are the rule. A typical double-extraction sequence of an aqueous solution by diethyl ether is diagramed in Figure. 3.5.

The original aqueous solution that has already been extracted once and drained into a flask is returned to the dirty, but empty, separatory funnel, along with a fresh portion of extraction solvent. If a second separatory funnel is available, the aqueous layer can be drained directly into it for the second extraction. (If the solvent forms a *lower* layer, it is drained off first. In this case, the original solution to be reextracted can simply remain in the separatory funnel, and fresh solvent added.) The second (and possibly a third) extraction is carried out just as the first one was. After the multiple extraction, the organic layers are combined in one flask. The aqueous layer (if that is the layer that will be discarded) should be labeled and saved until the entire experiment is finished.

B. Salting Out

If a compound has a low distribution coefficient between an organic solvent and water, one or more simple extractions will not remove much of the compound from the water.

The distribution coefficient of an organic compound between an organic solvent and water can be changed by adding sodium chloride to the water. (Other inorganic salts have the same effect as sodium chloride, but the

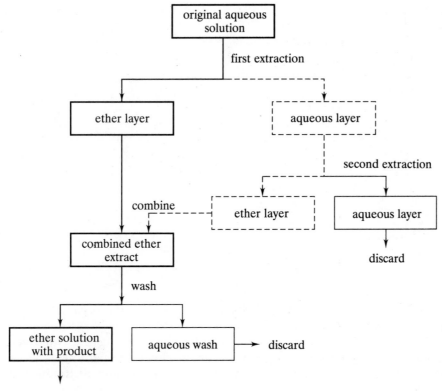

FIGURE 3.5 Diagram of a two-stage diethyl ether extraction of an aqueous solution. (The location of the organic product is indicated by the heavy black lines.)

latter is the least expensive salt available.) Organic compounds are less soluble in salt water than in plain water. Sometimes, the solubility difference is dramatic. Therefore, by simply dissolving sodium chloride in the water layer, we can increase the distribution of an organic compound in the organic solvent. This effect is commonly referred to as **salting out** the organic compound.

C. Emulsions

An **emulsion** is a suspension of one material in another that does not separate quickly by gravity. In liquid–liquid extractions, an emulsion refers to the suspension of one of the solutions in the other (see Figure 3.6). The result is that the two layers do not separate completely, a very annoying situation.

   If it is known that a particular extraction might lead to an emulsion, a few preventive steps might save time later. The *addition of sodium chloride to*

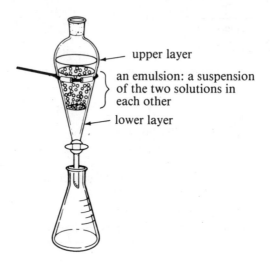

upper layer

an emulsion: a suspension
of the two solutions in
each other

lower layer

FIGURE 3.6   An emulsion in the separatory
funnel.

the aqueous layer decreases the solubility of water in the organic solvent,
and vice versa, and therefore may prevent an emulsion. *Swirling* the separatory funnel, instead of shaking it vigorously, when mixing the two layers
may also prevent an emulsion from forming. However, swirling is a less
efficient (and thus slower) way of reaching the equilibrium distribution of
solute.

Once an emulsion has formed, it can sometimes be broken up by one or
more of the following techniques:

1) Allow the separatory funnel to sit in an iron ring for a few minutes with
   periodic gentle swirling.
2) Add a few drops of saturated solution of aqueous sodium chloride with a
   dropper.
3) Squirt a few drops of 95% ethanol or a commercial antifoam agent on the
   emulsion "bubbles."

If none of the preceding techniques breaks up the emulsion, one of the
following techniques might work:

4) Filter the mixture with vacuum (which removes solid particles that help
   stabilize the emulsion), then proceed with the separation.
5) Draw off as much of the lower layer in the separatory funnel as possible,
   adding fresh solvent to the top layer remaining in the funnel to dilute it,
   and swirl gently.
6) Pour the mixture into a flask, cork the flask, and allow the mixture to sit
   overnight or until the next laboratory period. (Do not let the mixture
   stand for any length of time in the separatory funnel; the stopcock will
   eventually leak under the pressure of liquid.)

D. Washing

If the original solution was an aqueous one, the organic layers, separately or combined, may be *washed* (extracted with fresh water) to remove any water-soluble impurities. Use a volume of wash water about 10% that of the organic layer. A 5% sodium bicarbonate wash may be used to remove traces of acid from the organic layer. A sodium chloride wash (saturated aqueous solution) may be used to remove water from the organic solution.

For washing, place the organic solution in the separatory funnel along with the wash solution, and then shake and separate in the usual manner. Combine the aqueous washes with the original water solution. Save $NaHCO_3$ or NaCl washes separately. Label and save all extracts until the experiment is complete.

E. Drying and Removal of Solvent

No two liquids are completely immiscible. In any liquid–liquid extraction, the desired layer (usually the organic layer) contains some of the other solvent (usually water). Before the organic solvent is removed, the organic solution should be dried so that water will not contaminate the product. Specific procedures for drying organic solutions are discussed in Technique 4.

After an extraction has been completed and the extract dried, the solvent is removed from the organic compound. There are many ways to do this. Small amounts of solvent can be removed by simple evaporation or by boiling the solvent and using a stream of clean, dry air or vacuum to remove the vapors from the flask. (When heating such a mixture, take care that the residue in the flask does not become overheated.)

Large amounts of solvent should not be boiled away into the atmosphere, even through a fume hood; instead, the solvent should be distilled and collected (see Technique 5).

In some laboratories, you may have access to a *rotary evaporator*, a convenient vacuum apparatus for removing solvent quickly (see Figure 3.7). If your laboratory is equipped with one of these, your instructor will show you how to use it.

SAFETY NOTE Organic extraction solvents should be boiled or evaporated only in a fume hood. Flammable solvents should never be heated with a burner. A steam bath is the safest source of heat. A spark-free hot plate may also be used.

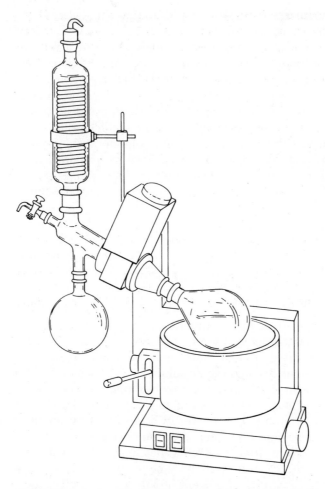

FIGURE 3.7   A typical roto-evaporator used for the removal of solvent under vacuum. (Courtesy of VWR Equipment Catalog. Reprinted by permission.)

· · · · · · · · · · ·
3.5   Chemically Active Extraction

The type of extraction procedure that we have been discussing can be considered a "passive" process—the extraction of a compound by virtue of its distribution between a pair of solvents. A less common, but very powerful, extraction technique is **chemically active extraction.** In this type of extraction, a compound is altered chemically to change its distribution between a pair of solvents. The most common method of changing the chemical structure is by using an acid–base reaction.

To illustrate how chemically active extraction works, let us consider a specific example. Assume we have a mixture of two compounds, a hydrocarbon and a carboxylic acid. Further assume that the hydrocarbon and the carboxylic acid are both soluble in diethyl ether and insoluble in water. These two compounds cannot be separated from one another by passive extraction. However, if the carboxylic acid is converted to an anion, which will be soluble in water but insoluble in diethyl ether, then we could effect a clean separation of the carboxylic acid (as its anion) from the hydrocarbon.

A mixture that is soluble in ether and insoluble in water:

a hydrocarbon   a carboxylic acid
neutral   acidic

If this mixture is treated with an aqueous solution of a base, such as sodium hydroxide solution, the carboxylic acid will react to form a water-soluble anion, but the hydrocarbon is neutral and will not react. It will remain water-insoluble and dissolved in the ether layer. By reaction, we have changed the carboxylic acid to an anion and thus have altered its distribution between water and ether.

Formation of the water-soluble salt:

water-insoluble;   an anion, water-soluble;
ether-soluble   ether-insoluble

By extraction of an ether solution containing a hydrocarbon and a carboxylic acid with an aqueous solution of NaOH, we can separate the two compounds. The hydrocarbon remains in the ether layer. The carboxylic acid reacts with the NaOH to form an anion, which then leaves the ether layer and dissolves in the water layer. The aqueous layer is separated from the ether layer using a separatory funnel and then acidified to regenerate the original carboxylic acid.

Regeneration of the carboxylic acid:

an anion   a carboxylic acid

The overall chemical extraction is given in Figure 3.8.

Carboxylic acids and phenols are weak acids. Amines are weak bases. Compounds belonging to these classes can often be separated from other

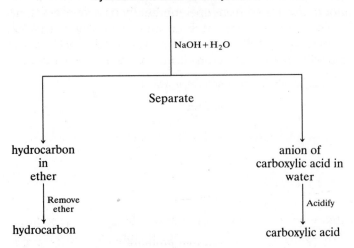

FIGURE 3.8   Separation of a carboxylic acid and a hydrocarbon by chemically active extraction.

organic compounds because they can be converted to anions by base (in the case of carboxylic acids and phenols) or to cations by acid (in the case of amines).

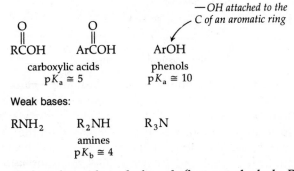

Carboxylic acids and phenols (but not alcohols, ROH) are strong enough acids to react with a strong aqueous base such as sodium hydroxide to yield water-soluble salts. In this form, these compounds can be extracted from neutral organic compounds.

$$\underset{\text{R\overset{O}{\overset{\|}{C}}OH}}{} + Na^+ + OH^- \longrightarrow \underset{\text{a water-soluble salt}}{\underline{RC\overset{O}{\overset{\|}{}}O^- + Na^+}} + H_2O$$

$$ArOH + Na^+ + OH^- \longrightarrow \underset{\text{a water-soluble salt}}{\underline{ArO^- + Na^+}} + H_2O$$

Carboxylic acids, with pK_a values typically around 5, undergo an acid–base reaction with the weak base sodium bicarbonate ($NaHCO_3$). A typical phenol, with a pK_a value of around 10, is only 1/100,000th as strong an acid as a carboxylic acid. Most phenols are too weakly acidic to undergo reaction with sodium bicarbonate. This difference in reactivity can be used to separate a mixture of a carboxylic acid and a phenol.

$$
\underset{\substack{\text{water-insoluble}}}{\overset{\overset{\displaystyle O}{\|}}{RCOH}} + Na^+ + HCO_3^- \longrightarrow \underset{\substack{\text{a water-soluble salt}}}{\underbrace{\overset{\overset{\displaystyle O}{\|}}{RCO^-} + Na^+}} + H_2O + CO_2\uparrow
$$

$$
\underset{\substack{\text{water-insoluble}}}{ArOH} + Na^+ + HCO_3^- \longrightarrow \text{no appreciable reaction}
$$

Amines are bases for the same reason that ammonia is a base: an amine contains a nitrogen atom with an unshared pair of electrons and thus can accept a proton. Treatment of an amine with aqueous acid yields a water-

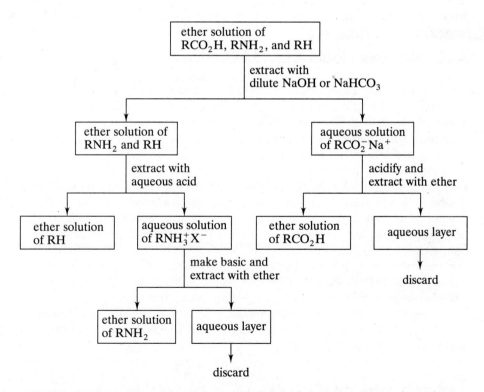

FIGURE 3.9 Separation of an organic acid (RCO_2H), an organic base (RNH_2), and a neutral organic compound (RH) by an acid–base extraction.

soluble salt that can be separated from water-insoluble organic compounds.

$$\overset{\displaystyle\frown}{R\ddot{N}H_2} + H^+ + Cl^- \longrightarrow \underbrace{\overset{\displaystyle H}{\overset{|}{R\overset{}{N}H_2^+}} + Cl^-}$$

RNH₂ — water-insoluble

RNH₂⁺ + Cl⁻ — a water-soluble salt

Because different compounds form water soluble salts under different conditions, the acid–base reactions that we have presented can form the basis of a number of types of chemically active extractions. In the laboratory, the acids and bases usually used to effect these reactions are as follows:

5%–10% aqueous HCl to extract an amine as a water-soluble salt
5% aqueous $NaHCO_3$ to extract RCO_2H as a water-soluble salt
5% aqueous NaOH to extract ArOH as a water-soluble salt

In any of these acid–base reactions, hydrocarbons, halogenated hydrocarbons, alcohols, and other neutral organic compounds are usually unaffected.

Figure 3.9 outlines the separation of an organic acid, an organic base, and a neutral compound by an acid–base extraction.

3.6 Other Extraction Techniques

A. Continuous Liquid–Liquid Extraction

The extraction of a compound with a low distribution coefficient between organic solvent and water would ordinarily require large amounts of solvent. The technique of continuous liquid–liquid extraction allows solvent to be recycled through the aqueous solution so that the compound can be completely extracted with a moderate amount of solvent.

Figure 3.10 illustrates a continuous-extraction apparatus that can be used with a lighter-than-water solvent. (Other types of extractors have been designed for heavier-than-water solvents.) The aqueous solution to be extracted is placed in a long tube with a sidearm. Solvent is placed in a distillation flask, as shown in the figure. When the solvent is distilled, its condensed vapors drip into a narrow glass tube with a fritted bottom. When the narrow tube is filled with solvent, small bubbles of solvent are forced through the frit to rise through the aqueous solution, extracting organic material as they rise. The organic solution above the water passes through the sidearm back to the distillation flask, where more solvent is distilled. As the desired organic material is extracted from the water, its concentration builds up in the distillation flask.

A continuous extraction of this sort requires several hours, or even days, but the operator is free to do other activities while the extraction is being carried out. When the extraction is complete, the organic extract is dried and the organic compound is isolated from solution.

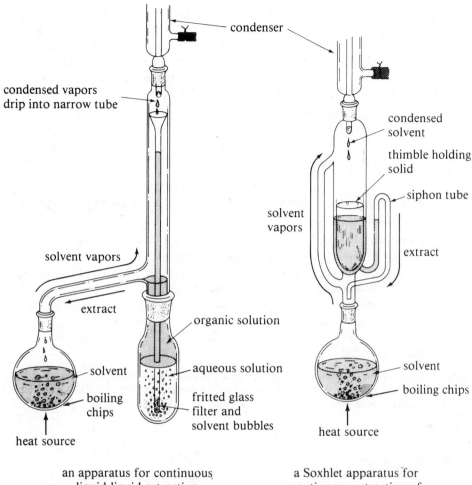

an apparatus for continuous
liquid-liquid extraction
using a solvent that is
lighter than water

a Soxhlet apparatus for
continuous extraction of
organic compounds
from a solid

FIGURE 3.10   Setups used for continuous extractions. (Clamps holding the flasks and tubes are not shown.)

For more details concerning continuous liquid–liquid extraction, consult the Suggested Readings at the end of this section, particularly *Technique of Organic Chemistry III.*

B. Continuous Solid–Liquid Extraction

For the continuous extraction of organic compounds from a solid mixture, a Soxhlet extraction apparatus (Figure 3.10) may be used. The solid material is placed in a porous cup, or thimble, made of thick, tough filter paper. The thimble is placed in the inner tube of the extractor. The apparatus allows hot solvent vapors to bypass the thimble through the sidearm, condense, and drip into the thimble. When the solvent fills the thimble, the organic extract siphons through the small siphon tube into the solvent distillation flask. The process is repeated automatically until the extraction is complete. Like continuous liquid–liquid extraction, several hours, even days, may be required for the extraction.

Suggested Readings

Morton, A. A. *Laboratory Techniques in Organic Chemistry*. New York: McGraw-Hill, 1938, Chapter 10.

Linstead, R. P., Elvidge, J. A., and Whalley, M. *A Course in Modern Techniques of Organic Chemistry*. London: Butterworths, 1955.

Evans, T. W. "The Calculation of the Maximum Results Obtainable by Extraction with Immiscible Solvents." *J. Chem. Ed.* **1936**, *13(11)*, 536.

Evans, T. W. "The Calculation of the Limiting Results Obtainable by Extraction with Partially Miscible Solvents." *J. Chem. Ed.* **1937**, *14*, 408.

Sharefkin, J. G., and Wolfe, J. M. "Efficiency in Extraction with Solvents." *J. Chem. Ed.* **1944**, *21(9)*, 449.

Craig, L. C., and Craig, D. "Extraction and Distribution." *Technique of Organic Chemistry III*. Weissberger, A., ed. New York: Interscience, 1950, p. 247. (Continuous liquid–liquid extraction.)

Problems

3.1 What is *wrong* with the following procedure?

The reaction mixture, consisting of NaBr and an ethanol solution of the product, is diluted with an equal volume of water, then extracted once with an equal volume of diethyl ether. The lower aqueous layer is discarded.

3.2 Suppose you added an additional 50 mL of water to a separatory funnel containing a compound distributed between 50 mL of ether and 50 mL of water. How will this addition affect (a) the distribution coefficient of the compound and (b) the actual distribution of the compound?

3.3 The pain reliever phenacetin is soluble in cold water to the extent of 1.0 g/1310 mL and soluble in diethyl ether to the extent of 1.0 g/90 mL.
   (a) Determine the approximate distribution coefficient for phenacetin in these two solvents.
   (b) If 50 mg of phenacetin were dissolved in 100 mL of water, how much ether would be required to extract 90% of the phenacetin in a single extraction?
   (c) What percent of the phenacetin would be extracted from the aqueous solution in (b) by two 25-mL portions of ether?

3.4 Complete the following equations. If no appreciable reaction occurs, write "no reaction."

(a) CH_3—⟨O⟩—OH + $NaOH$ $\xrightarrow{H_2O}$

(b) ⟨O⟩—CH_2CH_2OH + $NaOH$ $\xrightarrow{H_2O}$

(c) CH_3—⟨O⟩—OH + $NaHCO_3$ $\xrightarrow{H_2O}$

(d) $CH_3CH_2CH_2\overset{\displaystyle O}{\overset{\displaystyle \|}{C}}OH$ + $NaHCO_3$ $\xrightarrow{H_2O}$

(e) $CH_3NHCH_2CH_2NH_2$ + excess HCl $\xrightarrow{H_2O}$

3.5 Which of the following pairs of compounds *could* be separated by chemically active extraction? What reagent would you use?

(a) $CH_3CH_2\overset{\displaystyle O}{\overset{\displaystyle \|}{C}}OH$ $ClCH_2CH_2\overset{\displaystyle O}{\overset{\displaystyle \|}{C}}OH$

(b) $CH_3CH_2CH_2$—⟨O⟩—OH CH_3CH_2—⟨O⟩—CH_2OH

(c) ⟨⟩NH   ⟨⟩O

3.6 Draw a flow diagram for the separation of $CH_3CH_2CH_2Br$, $(CH_3CH_2CH_2)_3N$, and $CH_3CH_2CO_2H$.

3.7 At a pH of 8, what percent of phenol ($pK_a = 10.00$) exists as an anion?

⟨O⟩—OH

phenol

3.8 An aqueous solution containing 5.0 g of solute in 100 mL is extracted with three 25-mL portions of diethyl ether. What is the total amount of solute that will be extracted by the ether in each of the following cases?
(a) Distribution coefficient (ether–water), $K = 0.10$
(b) $K = 1.0$
(c) $K = 10$

3.9 If the compound in Problem 3.8(b) were extracted with three 50-mL portions of diethyl ether, how much would be extracted?

3.10 Diagram a two-stage dichloromethane ($d = 1.3$) extraction of an aqueous solution, showing how this procedure differs from the diethyl ether extraction diagramed in Figure 3.5.

3.11 Suppose that you wish to extract 10 g of benzoic acid from an ether solution with aqueous $NaHCO_3$.
(a) What is the minimum amount of $NaHCO_3$ that must be added to 50 mL of water to carry out the extraction?
(b) What is the minimum amount of concd HCl (37%) that would be needed to convert the sodium benzoate back to its free acid?

TECHNIQUE 4

Drying
Organic Solutions

Most organic solutions are dried after an aqueous extraction and before solvent is removed by distillation or by use of a roto-evaporator. Various methods for removing water from wet organic solutions are available. The method chosen depends on a number of factors, including the organic compound in solution, the solvent, the degree of wetness, and the degree of dryness desired.

4.1 Extraction with Aqueous Sodium Chloride

A saturated aqueous solution of sodium chloride is an inexpensive drying agent that will remove the bulk of water from a wet organic solution. A final extraction using a separator funnel with a saturated NaCl solution is especially valuable when the organic extraction solvent is diethyl ether, which can contain 1.2% water at 20°. When most of the water is removed by a saturated NaCl wash, the solution can be further dried with a solid inorganic drying agent.

4.2 Solid Inorganic Drying Agents

Solid drying agents are commonly used to remove the last traces of water from organic solutions. These drying agents are generally anhydrous inorganic salts (insoluble in organic liquids) that absorb water and are converted to hydrated salts. Molecular sieves are aluminosilicates that contain cavities

TABLE 4.1 Some drying agents for organic solutions

Name	Formula	Speed	Practical capacity[a]	Comments
molecular sieve 4A (powder)[b]	—	very fast	very high	neutral and easy to use, but expensive
calcium chloride	$CaCl_2$	fast	high	forms complexes with O and N compounds, such as alcohols, amines, ketones, and carboxylic acids; may contain CaO as an impurity
magnesium sulfate	$MgSO_4$	moderate	high	best filtered, not decanted, because of fine particle size
potassium carbonate	K_2CO_3	moderate	moderate	basic—reacts with acidic compounds such as phenols and carboxylic acids
calcium sulfate	$CaSO_4$	fast	low	neutral; available with indicator
sodium sulfate	Na_2SO_4	fast	low	neutral; easy to use
saturated sodium chloride solution	$NaCl + H_2O$	very fast	low	used to remove the bulk of water from an organic solution so that less solid drying agent is needed

[a]Based on the quantity of water removed from wet ether per gram of drying agent.
[b]Molecular sieves are also available as beads, which are not as fast or efficient as the powdered form.

that can trap molecules of certain sizes and shapes. Some types of molecular sieves are excellent drying agents.

A moderate amount of drying agent (about 1–5 g, or just enough to cover the bottom of the flask) is sufficient for drying most solutions. Add the drying agent, cork the flask, and swirl the contents of the flask to speed the drying. If the drying agent becomes wet looking or clumped, filter or decant the solution into a clean, dry flask and add a fresh portion of drying agent. Finally, remove the hydrated salts by filtering. The hydrated salts can also be removed by carefully decanting the organic solution into a clean flask. The residual organic solution can be transferred by using a disposable pipet.

Table 4.1 lists a few common drying agents and comments about their use. Most drying agents are reasonably swift (15 minutes) in removing the bulk of the water from an organic solution; however, most are quite slow in trapping the last vestiges of water. For this reason, an overnight drying is always preferable to a 15-minute drying. If a trace of water is tolerable, however, the 15-minute drying may allow you to continue the experiment and save laboratory time.

The amount of water a drying agent can remove from a wet solvent depends on the solvent and the structure and stability of the hydrate. For example, calcium sulfate, which forms the hydrate $CaSO_4 \cdot \frac{1}{2} H_2O$, cannot remove as much water as can calcium chloride, which can form hydrates containing 1–6 moles of water per mole of $CaCl_2$.

Your choice of drying agent will depend on the compound you are drying, the drying time available, the degree of dryness necessary, and the expense. For everyday classroom use, anhydrous calcium chloride, magnesium sulfate, and sodium sulfate are the most useful. Table 4.1 lists the limitations of these drying agents. Note that calcium chloride should not be used for drying compounds containing unshared valence electrons, such as alcohols and amines.

In special cases, drying agents that actually react with water are used. For example, diethyl ether is often dried over sodium metal. Very wet ether cannot be dried this way because of the violent reaction of sodium with water $(2\,Na + 2\,H_2O \rightarrow 2\,NaOH + H_2)$. Ether solutions of organic compounds are also not dried over sodium. Calcium hydride is another example of a drying agent that reacts quite vigorously with water $(CaH_2 + H_2O \rightarrow CaO + 2\,H_2)$.

4.3 Azeotropic Drying

Some solvents form **low-boiling azeotropes** with water—that is, when they are distilled, they produce a mixed distillate of constant composition with a lower boiling point than that of either water or the pure solvent (see also Section 5.1D). For example, benzene (bp 80.1°) and water (bp 100.0°) form an

TABLE 4.2 Composition and boiling points of some low-boiling binary azeotropes containing water

Composition	bp of azeotrope (°C)
91% benzene (bp 80.1°) 9% water (bp 100.0°)	69.4
96% carbon tetrachloride (CCl_4; bp 76.8°) 4% water	66.8
97% chloroform ($CHCl_3$; bp 61.7°) 3% water	56.3
99% dichloromethane (CH_2Cl_2; bp 40°) 1% water	38.8

azeotrope composed of 91% benzene and 9% water that boils at 69.4°. Table 4.2 lists some solvents that form low-boiling azeotropes with water.

Solutions in these solvents may be dried by simply distilling the azeotropic mixture until the distillate is clear and no longer a two-phase mixture of water plus solvent. Once the azeotrope is distilled, the residual solution contains no more water.

A Dean-Stark trap is a device used to remove water from a reaction mixture or solution by azeotropic distillation. To use a Dean-Stark trap, assemble the device as shown in Figure 4.1. The organic solution to be dried (or reaction mixture from which water is to be removed) is placed in the round-bottom flask. The organic solvent must be one that forms azeotropes with water. In addition, the solvent used for the azeotropic drying must be less dense than water. Benzene is a suitable solvent for use in a Dean-Stark apparatus.

The solution to be dried is boiled, causing the azeotrope to distill and be condensed in the condenser. Upon condensation, the azeotrope separates into two layers—the organic solvent layer and the water layer. Both layers drain into the Dean-Stark trap. The heavier water sinks to the bottom of the trap. The lighter solvent floats on top of the water and, when it fills the trap, drains back into the flask.

The Barrett receiver (see Figure 4.1) is a similar to a Dean-Stark trap. The difference is that the Barrett receiver contains a stopcock at the bottom of the water trap. This allows the periodic removal of water as it accumulates. The major disadvantage of the Barrett receiver is that the stopcock tends to leak when used for an extended period of time.

Suggested Readings

Morton, A. A. *Laboratory Techniques in Organic Chemistry.* New York: McGraw-Hill, 1938, Chapter 1.

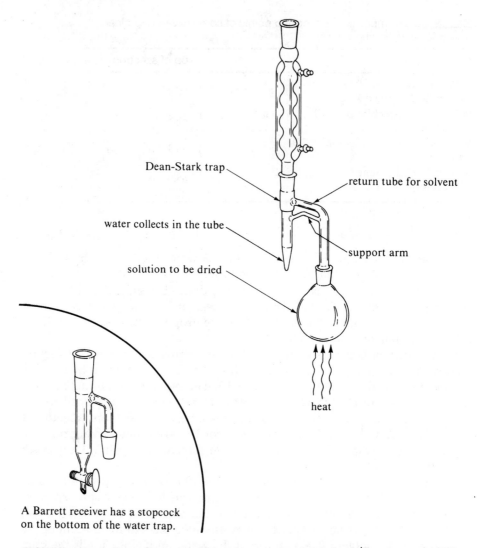

Dean-Stark trap

return tube for solvent

water collects in the tube

support arm

solution to be dried

heat

A Barrett receiver has a stopcock
on the bottom of the water trap.

FIGURE 4.1 A Dean-Stark apparatus for azeotropic drying. (Courtesy of VWR Equipment Catalog. Reprinted by permission.)

Burfield, D. R., Lee, K., and Smithers, R. H. "Desiccant Efficiency in Solvent Drying." *J. Org. Chem.* **1977,** *42(8),* 3060.

Burfield, D. R., and Smithers, R. H. "Drying of Grossly Wet Ether Extracts." *J. Chem. Ed.* **1982,** *59(8),* 703.

Burfield, D. R., and Smithers, R. H. "Desiccant Efficiency in Solvent and Regent Drying. 7. Alcohols." *J. Org. Chem.* **1983,** *48(14),* 2420.

Problems

4.1 Which drying agent(s) in Table 4.1 could be used to dry an ether solution of each of the following compounds?

(a) $CH_3CH_2CH_2CH_2OH$ (b) $CH_3CH_2CH_2CH_2Br$

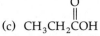
(c) CH_3CH_2COH (d) $CH_3CH_2CH_2CH_2NH_2$

4.2 Why must a solid drying agent be removed from a solution before the solution is distilled?

4.3 Sodium sulfate can form a decahydrate, and yet anhydrous sodium sulfate is a less efficient drying agent than calcium sulfate. Suggest a reason for this behavior.

4.4 Because anhydrous sodium sulfate is a relatively inefficient drying agent, a student uses twice as much of this salt as recommended. What are the advantages and disadvantages of adding extra drying agent?

TECHNIQUE 5

Simple Distillation

....

Distillation is a general technique used for removing a solvent, purifying a liquid, or separating the components of a liquid mixture. In distillation, a liquid is vaporized by boiling, then condensed back to a liquid, called the **distillate** or **condensate,** and collected in a separate flask (the **receiving flask**). In an ideal situation, a low-boiling component can be collected in one flask, and then a higher-boiling component can be collected in another flask, while the highest-boiling components remain in the original distillation flask as the **residue.**

....
5.1 Characteristics of Distillation

A. The Boiling Point

The **boiling point** of a liquid is defined as the temperature at which its vapor pressure equals atmospheric pressure, and it is characterized by vigorous bubbling and churning of the liquid as it vaporizes. Even at a constant atmospheric pressure, the temperature of a boiling liquid is seldom reproducible because of the presence of impurities and the possibility of superheating. The actual temperature of a boiling liquid is always *higher* than the boiling point of the material distilling. Therefore, the boiling point is measured above the surface of the liquid, where vapor and liquid are in equilibrium.

Figure 5.7 in Section 5.2 shows the apparatus for a **simple distillation.** When the liquid in the distillation flask boils, vapor rises to the top of the flask, through the distillation head, past the thermometer, and out the sidearm into the condenser. Some of the vapor condenses on the walls of

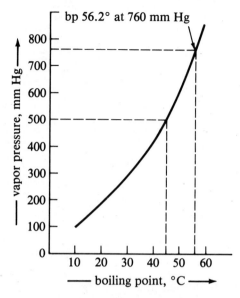

FIGURE 5.1 Vapor pressure–boiling-point diagram for acetone, $(CH_3)_2C{=}O$.

the flask and head and on the thermometer. As long as vapor is flowing out the sidearm and the thermometer bulb is immersed in the vapor, the condensed liquid on the tip of the thermometer is in equilibrium with the vapor, and an accurate boiling point can be determined. Like the melting point, a boiling point is usually reported as a range. A pure compound exhibits a range of 1°–2° or less (but not all liquids with constant boiling points are single pure compounds; see Section 5.1D).

The boiling point is often reported at a particular pressure—for example, bp 83.5°–84.5° (752 mm Hg). The reason is that boiling points vary significantly with changes in pressure. Figure 5.1 is a graph of the vapor pressure of acetone versus its boiling point. Because a liquid boils when its vapor pressure equals the pressure above it, the term *vapor pressure* in the figure actually means "applied pressure". Note that acetone boils at about 56° at 1.0 atmosphere of pressure (760 mm Hg or 760 torr). At a 500-mm Hg pressure, the boiling point of acetone is only about 45°.

B. Distillation of a Single Volatile Liquid

When a solvent containing a nonvolatile component is removed from a solution by heating, or when a reasonably pure liquid is distilled, the observed temperature rises rapidly to the boiling point. Once the distillation apparatus has reached thermal equilibrium, the boiling point remains relatively constant, not changing more than 1°–2° during the course of the

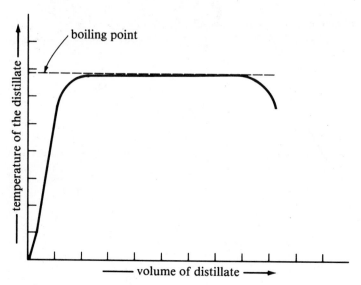

FIGURE 5.2   Volume versus temperature of the distillate in the distillation of a pure liquid.

distillation. A significant drop in temperature is the signal that the distillation of the solvent or the pure liquid is complete (see Figure 5.2).

C. Distillation of Mixtures

Because most distillations are performed with mixtures, let us consider the theory of distillation of an ideal (noninteracting) mixture of two miscible liquids A and B, where A is the lower-boiling component. If the difference in boiling points between A and B is large (100° or more), then the distillation temperature will rise to the boiling point of compound A, the lower-boiling component, and remain constant while A is distilling. When almost all of compound A has distilled, the temperature will begin to rise rapidly toward the boiling point of compound B, the higher-boiling component. Once it reaches the boiling point of B, it will remain constant while B is distilling. Unfortunately, this type of distillation is rarely encountered in laboratory practice, because separations usually involve mixtures of compounds with boiling points closer together than 100°.

A more common experience is that the distillation temperature rises more or less steadily during the distillation because *mixtures* of A and B distill. The first, lower-boiling portions of the distillate contain more of A than of B, but this distillate will not be pure A. Similarly, the last, higher-boiling portions of the distillate will be predominantly B but will also contain some A. Although enrichment of A and B in the first and last portions of the distillate can be accomplished, neither pure A nor pure B will be obtained. The reason is that B has a significant vapor pressure, even at temperatures below its boiling point. The vapor pressure of B in a boiling mixture of A and B is a function of a number of factors. To discuss these, we must consider Dalton's law and Raoult's law.

Dalton's law and Raoult's law A liquid boils when its vapor pressure equals the atmospheric pressure. **Dalton's law of partial pressures** states that the total pressure of a gas is the sum of the partial pressures of its individual components. Thus, the total vapor pressure of a liquid (P_{total}) is the sum of the partial vapor pressures of its components. In our example, the mixture of A and B boils when the sum of the two partial pressures (P_A and P_B) equals the atmospheric pressure.

$$P_{total} = P_A + P_B$$

The **mole fraction X,** a measure of concentration, is the number of moles of one particular component in a mixture divided by the total number of moles (all components) present.

$$X_A = \frac{\text{moles of A}}{\text{moles of A + moles of B}}$$

$$X_B = \frac{\text{moles of B}}{\text{moles of A + moles of B}}$$

where X_A and X_B are the mole fractions of A and B in the mixture

Raoult's law states that, at a given temperature and pressure, the partial vapor pressure of a compound in an ideal solution is equal to the vapor pressure of that pure compound multiplied by its mole fraction in the liquid. By Raoult's law, if P_A^0 represents the vapor pressure of *pure* A at a specific temperature, then the partial vapor pressure of A (P_A) in a solution is equal to $X_A P_A^0$. Similarly, the partial vapor pressure of B (P_B) at that temperature is $X_B P_B^0$.

For the liquid mixture:

$$P_A = X_A P_A^0 \qquad \text{and} \qquad P_B = X_B P_B^0$$

And because $P_{total} = P_A + P_B$, then

$$P_{total} = X_A P_A^0 + X_B P_B^0$$

From Dalton's law and Raoult's law, we conclude that the total vapor pressure of the liquid is a function of the vapor pressures of the pure components and their mole fractions in the mixture.

Composition of the liquid Distillation is a dynamic process. Vapor is removed as condensed distillate, while more vapor is generated by the boiling liquid. During the course of distillation, the liquid mixture of A and B becomes progressively poorer in the lower-boiling component A and richer

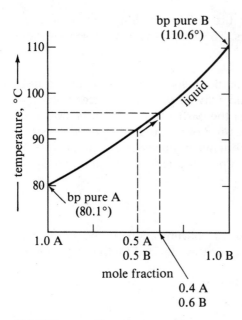

FIGURE 5.3 The boiling point of an ideal binary liquid solution versus mole fractions of components in the liquid. (The arrow on the curve shows the direction of change in temperature and composition as the distillation proceeds.)

in the higher-boiling component B. The combination of Dalton's law and Raoult's law show this mathematically.

To illustrate how the composition changes during a distillation, let us consider what happens to a mixture of A (bp 80.1°) and B (bp 110.6°) in a molar mixture of $1:1$ ($X_A = 0.50$ and $X_B = 0.50$). Compound A boils at a lower temperature than does compound B; therefore, P_A^0 is larger than P_B^0. For this reason, compound A initially contributes a larger partial vapor pressure ($P_A = X_A P_A^0$). At the start of the distillation, the vapor from the boiling mixture contains more A than B. As the distillation proceeds, the boiling liquid contains progressively less A: X_A decreases and X_B increases.

At the start, greater value for P_A^0 means more A distills.

$$P_{total} = X_A P_A^0 + X_B P_B^0$$

As the distillation progresses and A is removed, a greater value for X_B means an increasing amount of B distills.

Because P_B^0 is lower than P_A^0, the temperature of the liquid must be increased to maintain a boil as the liquid becomes richer in B. Figure 5.3 is a plot of the relative concentrations of A and B versus the temperature of the

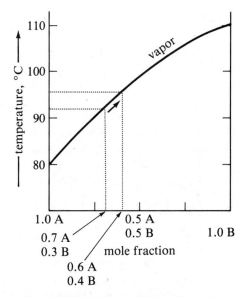

FIGURE 5.4 The boiling point of an ideal binary liquid solution versus mole fractions of components in the vapor.

boiling liquid. In a distillation, we would start at one point on the curve and move to the right.

At the start of the distillation of a 1:1 mixture of A and B, the vapor contains a greater mole fraction of A because P_A is greater than P_B. As the distillation proceeds, P_A decreases because of the decreasing amount of A in the liquid, and thus X'_A decreases.

Composition of the vapor The mole fraction of a compound such as A in the vapor above a boiling mixture (not in the boiling liquid itself) is equal to the ratio of its partial vapor pressure (P_A) to the total pressure.

For the vapor:

$$X'_A = \frac{P_A}{P_{total}} \quad \text{and} \quad X'_B = \frac{P_B}{P_{total}}$$

where X'_A and X'_B are the mole fractions of A and B in the vapor.

Figure 5.4 is a plot of the vapor composition versus the temperature of the vapor. The temperature of the vapor gradually rises as the amount of B in the vapor increases.

Relating liquid and vapor compositions Generally, the curves for liquid and vapor compositions versus temperature are combined into a single diagram called a *boiling-point–composition diagram*, or *phase diagram*, as shown in Figure 5.5 (a combination of the two curves previously presented in

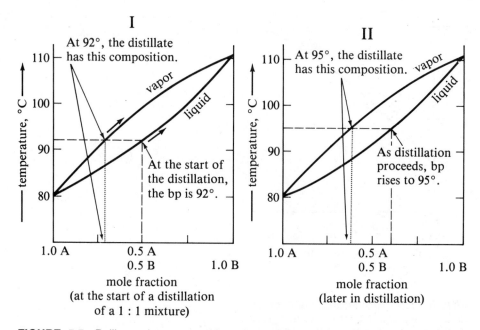

FIGURE 5.5 Boiling-point–composition diagram, or phase diagram, for an ideal binary solution. (A distillation of a mixture of A and B would begin at the liquid of that composition and progress from there to the right along the lower curve. The composition of the vapor at each temperature is read from the upper curve.)

Figures 5.3 and 5.4). In Figure 5.5, we show the same diagram twice to show what happens in the distillation of a 1 : 1 mixture of A and B. From the first diagram, we see that a 1 : 1 mixture of A and B boils at 92°, and that the vapor at 92° contains approximately 70 mole % A and 30 mole % B (mole fractions 0.7 and 0.3, respectively).

As the distillation continues, more A is distilled than B; therefore, the liquid contains a progressively larger percentage of B. When the boiling liquid contains 40% A and 60% B (mole fractions 0.40 and 0.60, respectively), its temperature is 95°. At 95°, however, the vapor contains 60% A–40% B (mole fractions 0.6 and 0.4). These compositions are marked in the second diagram in Figure 5.5. As the distillation continues, we move farther to the right on each curve. The net results of the changing compositions are (1) a steady increase in the boiling point, and (2) a distillate containing progressively less A and more B.

In our example, a distillate of pure A could never be obtained. However, the early portion of the distillate (containing 80% or 90% A, for example) could be redistilled, yielding somewhat purer A at the start of the distillation. Then, *this* distillate could be redistilled, yielding even purer A. Because these distillations would be tedious and time-consuming, *fractional distillation* was developed. We will discuss fractional distillation as *Technique* 6.

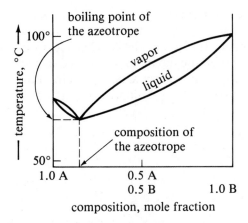

FIGURE 5.6   Phase diagram for a typical binary mixture that forms a low-boiling azeotrope.

D. Azeotropes

Because of intermolecular interactions, such as hydrogen bonding, many binary mixtures do not follow Raoult's law and do not have idealized phase diagrams, such as the ones shown in Figure 5.5. One example is a pair of liquids that forms an **azeotrope,** which is a mixture that distills at a constant boiling point and with a constant composition. The phase diagram for one type of azeotrope is depicted in Figure 5.6. The low point on the graph represents the azeotropic mixture, which boils at a constant boiling point, just as a pure compound does. Note that the boiling point of the azeotrope shown in Figure 5.6 is *lower* than that of either pure component. Because the boiling point is lower, the azeotrope will distill *before* a component present in excess. The excess component will not distill as a pure compound until the azeotrope has completely distilled.

Although most azeotropes have boiling points lower than those of their components, some azeotropes have boiling points intermediate between those of their components, while other azeotropes have higher boiling points than those of the components. For example, the azeotrope of α-bromotoluene (bp 183.7°) and *n*-octanol (bp 195°) boils at an intermediate temperature of 184.1°, and the azeotrope of acetone (bp 56°) and chloroform (bp 61°) boils at a higher temperature of 64.7°. Azeotropes of two, three, or even more components are also well known.

A common azeotropic mixture encountered in the laboratory is the binary azeotrope of ethanol (CH_3CH_2OH) and water (for others, see Table 4.2). For use in beverages, ethanol is made by the fermentation of aqueous sugars and starches. The fermentation of the carbohydrate mixture stops when the alcohol content of the mixture reaches 12%–14% by volume because ethanol at this concentration inhibits the action of the enzymes. To yield a higher concentration of alcohol, the fermentation mixture is distilled. Ethanol (bp 78.3°) and water (bp 100°) form a low-boiling azeotrope (bp 78.15°); therefore, this is the first major fraction to distill. This azeotrope contains 95% ethanol and 5% water. After all the ethanol has distilled as part of the azeotrope, the remaining water distills at its normal boiling point. Pure ethanol cannot be distilled from an ethanol–water solution that contains 5% or more water.

5.2 Steps in a Simple Distillation

The apparatus for a simple distillation is shown in Figure 5.7. Study this figure carefully, noting the placement and clamping of the distillation flask, the distillation head, and the condenser. Note that water flows into the bottom of the condenser's cooling jacket and out the top. If the water inlet were at the top, the condenser would not fill. Also, note the placement of the thermometer bulb, *just below the level of the sidearm* of the distillation head. If the bulb were placed higher than this position, it would not be in the vapor path and consequently would show an erroneously low reading for the boiling point.

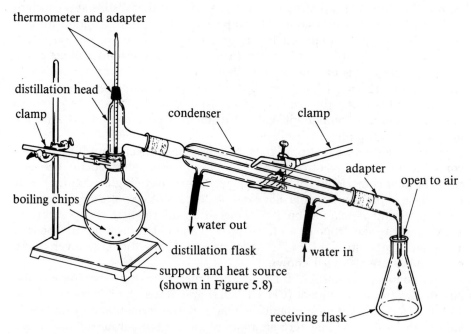

FIGURE 5.7 The apparatus for a simple distillation.

SAFETY NOTE   Distillation of noxious or toxic substances should always be carried out in a fume hood. Special precautions should also be taken with distillations of highly flammable substances, such as most solvents. Never use a burner in these cases, and avoid allowing an excess of uncondensed vapors to flow into the room.

1) The distillation flask.   Use only a round-bottom flask, never an Erlenmeyer flask, for distillation. The flask should be large enough that the material to be distilled fills $\frac{1}{2} - \frac{2}{3}$ of its volume. If the flask is overly large, a substantial amount of distillate will be lost as vapor filling the flask at the end of the distillation. If the flask is too small, boiling material may foam, splash, or boil up into the distillation head, thus ruining the separation.

Grease the ground-glass joint of the flask lightly; then securely clamp the flask to a ring stand or rack. Before adding liquid, support the bottom of the flask with a heating mantle on an iron ring (see Figure 5.8). A soft heating mantle should fit snugly around the flask. A hard mantle must fit, or be slightly larger than, the flask. Pour the liquid into the distillation flask, using a funnel with a stem to prevent the liquid from contaminating the ground-glass joint. Finally, add two or three boiling chips. (CAUTION: Never add boiling chips to a hot liquid!)

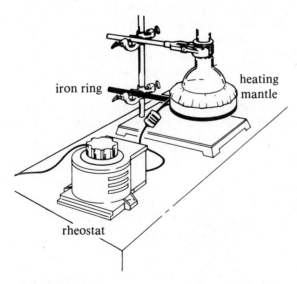

iron ring

heating mantle

rheostat

FIGURE 5.8   Heat source and support for a distillation flask.

2) The distillation head.   Grease the ground-glass joints of the distillation head lightly and place the head on the flask, rotating the joints to disperse the grease. It is usually not necessary to clamp the head. (Do *not* attach the thermometer at this time.)

3) The condenser.   Grease the ground-glass joints of the condenser lightly and attach rubber tubing firmly to the jacket inlet and outlet (which should *not* be greased). A strong clamp (oversized, if available) is needed to hold the condenser in place. The weight and angle of the condenser will tend to pull it away from the distillation head; therefore, check the tightness of this joint frequently before and during a distillation. Spring clamps can be used for holding these joints together but not for supporting weight.

Attach the rubber tubing from the lower end of the condenser to an adapter on the water faucet. Place the tubing from the upper end of the condenser in the sink or drainage trough. Turn on the water cautiously. Water should fill the condenser and gently flow, not drip, from the outlet tubing; however, a forceful flow of water will cause the tubing to pop off the condenser. To help prevent this, twist short pieces of wire around the tubing on the condenser inlet and outlet. Because water pressure varies and faucets tend to tighten gradually, check the flow of water frequently during the distillation.

4) The adapter.   The adapter directs the flow of distillate plus uncondensed vapors into the receiving flask. Figure 5.7 shows the usual type of adapter, but a vacuum adapter (Figure 5.9) may also be used. If desired, a piece of rubber tubing attached to the vacuum adapter can be used to carry fumes to the floor. (Rubber tubing is no substitute for a fume hood, however.) Whichever type of adapter is used, grease its joint lightly before attaching it to the condenser. A rubber band may be used to secure the adapter, as shown in Figure 5.9. A plastic clip can be used in place of a rubber band.

5) The receiving flask.   Almost any container can be used as a receiver, as long as it is large enough to receive the expected quantity of

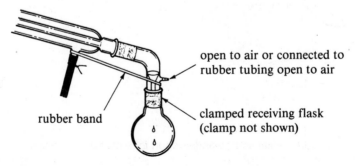

rubber band

open to air or connected to
rubber tubing open to air

clamped receiving flask
(clamp not shown)

FIGURE 5.9   A vacuum adapter.

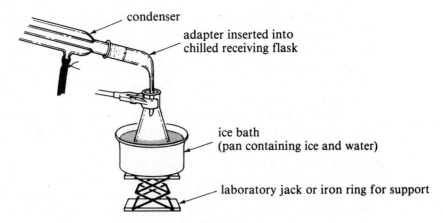

FIGURE 5.10   An ice-chilled distillation receiver.

distillate. An Erlenmeyer flask is recommended. A beaker is not recommended because its wide top allows vapors and splashes to escape and allows dirt to get into the distillate. Either set the receiving flask on the benchtop or clamp it in place. (It is not good practice to prop up a receiving flask on a stack of books.) If you are collecting several fractions, prepare a series of clean, dry, tared (weighed empty) flasks. If the volume, rather than the weight, of distillate is to be determined, you may use a clean, dry graduated cylinder as the receiver. A round-bottom flask with a ground-glass joint is also a good receiver. This type of flask will fit onto a vacuum adapter, but it must be clamped in place.

If a volatile compound is being distilled, the receiver can be chilled in an ice bath to minimize loss by vaporization (Figure 5.10). A *slow* distillation with an efficient condenser (keeping a moderately fast water flow at all times) also helps minimize loss of the compound.

SAFETY NOTE   When distilling at atmospheric pressure, *always* leave the apparatus open to the air at the adapter–receiver end. If you attempt a distillation with a closed system, the pressure buildup inside the apparatus may cause it to explode.

6) The thermometer.   Attach the thermometer last (and remove it first), because thermometers are expensive and easily broken. The easiest type of thermometer to insert is one with a ground-glass joint that fits a joint at the top of the distillation head. Neoprene adapters are available for attaching ordinary thermometers. Alternatively, a short piece of rubber tubing used as a sleeve can be used to hold the thermometer in place. A one-hole rubber stopper is not recommended because the hot vapors and condensate of the distilling liquid may dissolve some of the rubber, which will discolor the distillate. When attaching the thermometer, be sure to place the bulb just below the level of the sidearm, as shown in Figure 5.7, so that the bulb will be completely immersed in vapors of the distillate.

7) The actual distillation.   Before proceeding, check the water flow through the condenser and make sure that all ground-glass joints are snug. Plug the heating mantle into a rheostat; then plug the rheostat into the wall socket.

Slowly heat the mixture in the distilling flask to a gentle boil. You will then see the **reflux level** (the ring of condensate, or upper level of vapor condensing and running back into the flask) rise up the walls of the flask to the thermometer and sidearm. At this time, the temperature reading on the thermometer will rise rapidly until it registers the *initial boiling point,* which should be recorded. The vapors and condensate will pass through the sidearm and into the condenser, where most of the vapor will condense to liquid, and will finally drip from the adapter into the receiving flask.

The proper *rate of distillation* is one drop of distillate every 1–2 seconds. This rate is achieved by controlling the amount of heat supplied to the distillation flask. A slow rate means that not enough vapor is reaching the thermometer to give an accurate boiling point. A rapid rate results in uncondensed vapor being carried through the condenser and into the room. A rapid rate can also result in poor separation of components. It is generally necessary to increase the amount of heat applied to the distillation flask (by increasing the rheostat setting) during the course of a distillation. For distillations above 100°, insulate the head with crumpled aluminum foil to avoid temperature fluctuations.

8) Collecting the fractions.   Volatile impurities are the first compounds to distill. This first fraction, called the **fore-run,** is generally collected separately. When the temperature has risen to the desired level and has been recorded, place a fresh receiver under the adapter to collect the main fraction. In some cases, the main fraction can be collected in a single receiver. In other cases, it should be collected as a series of smaller fractions. Each time you change a receiver, note the temperature reading and record the boiling range of the fraction. After checking the purity of a group of fractions, you may want to combine some of these fractions later. Figure 5.11 shows typical notebook records of distillations collected in several fractions.

Impurities that are higher boiling than the desired material are generally not distilled but are left in the distillation flask as the residue. If higher-boiling impurities are present in large quantities, the temperature may rise from the desired level as the impurities begin to distill. However, the temperature frequently *drops* after the main fractions have distilled. This happens because not enough vapor and condensate are present in the head to keep the thermometer bulb hot. If the temperature drops at the end of a

Date: 10/15/83

	tare weight	Bp	gross weight	net weight
Fraction 1	25.93g	35-98°	30.32g	4.39 g
2	25.64	98-100	35.21	9.57
3	25.99	100-100	37.82	11.83
4	26.02	100-101	36.59	10.57
5	25.24	101-102	31.01	5.77
6	26.15	102-110	28.76	2.61
residue	—	—	—	approx. 3g

Date: 10/17/83

	Bp °C (749 mm)	volume, mL	n_D^{20}
Fraction 1	50-110	0.5	1.3372
2	110-112	1.0	1.4952
3	112-114	15.5	1.4960
4	114-125	6.0	1.4982
residue	—	2.0	1.5047

FIGURE 5.11 Typical notebook records of distillations.

distillation, the last temperature to record is the *highest* temperature, before the drop occurred.

At the conclusion of a distillation, remove the heat source. Turn off a heating mantle and lower it from the flask immediately. Allow the entire apparatus to cool before dismantling it.

SAFETY NOTE Never carry out a distillation to dryness, but always leave a small amount of residue in the distillation flask. The boiling residue will prevent the flask from overheating and breaking and will also prevent the formation of pyrolytic tars, which are difficult to wash out. Never dismantle a hot distillation apparatus — the hot residue or vapors might ignite when exposed to air.

Suggested Readings

Coulson, E. A., and Herington, E. F. G. *Laboratory Distillation Practice*. New York: Interscience Publishers, 1958.

Egloff, G., and Lowry, C. D. "Distillation as an Alchemical Art." *J. Chem. Ed.* **1930**, 7(9), 2063.

Houtman, J. P. W., and Husain, A. "Batch Distillation." *J. Chem. Ed.* **1955**, 32(10), 529.

Liebmann, A. J. "History of Distillation." *J. Chem. Ed.* **1956**, 33(4), 166.

Linstead, R. P., Elvidge, J. A., and Whalley, M. *A Course in Modern Techniques of Organic Chemistry*. London: Butterworths, 1955.

Problems

5.1   A mixture of two miscible liquids with widely different boiling points is distilled. The temperature of the distilling liquid is observed to plateau and then drop before rising again. Explain the temperature drop.

5.2   If a mixture is distilled rapidly, the separation of its components is poorer than if the mixture is distilled slowly. Explain.

5.3   From the graph in Figure 5.1 (page 95), estimate the following.
(a) The boiling point of acetone at 1000 mm Hg
(b) The vapor pressure of acetone at room temperature (23°C)
(c) The percent of the total pressure contributed by acetone above an open beaker of this compound at room temperature (23°) and 760 mm

5.4   A 50% aqueous solution of ethanol (50 mL total) is distilled and collected in 10-mL fractions. Predict the boiling range of each fraction.

5.5   Calculate the mole fraction of each compound in the following mixtures.
(a) 95.0 g CH_3CH_2OH and 5.0 g H_2O
(b) 10.0 g CH_3OH, 10.0 g CH_3CH_2OH, and 10.0 g $CH_3CH_2CH_2OH$

5.6   (a) Using the graph in Figure 5.5, determine the composition of the liquid solution boiling at 100°.
(b) What is the composition of the vapor being given off?

5.7   A mixture of ideal miscible liquids C and D is distilled at 760 mm Hg pressure. At the start of the distillation, the mole percent of C in the mixture is 90.0, while that of D is 10.0. The vapor is condensed and found to contain 15.0 mole percent of C and 85.0 mole percent of D. Calculate the following.
(a) The partial vapor pressures (P) of C and D in this mixture
(b) The vapor pressures (P^0) of pure C and D

5.8   Given the following mole fractions and vapor pressures for miscible liquids E and F, calculate the composition (in mole percent) of the vapor from a distilling ideal binary solution at 150° and 760 mm Hg.

For the solution:

$$X_E = 0.40 \qquad\qquad X_F = 0.60$$

$$P_E^0 = 1710 \text{ mm Hg} \qquad P_F^0 = 127 \text{ mm Hg}$$

TECHNIQUE 6

Fractional Distillation

....

The technique of **fractional distillation** differs from simple distillation in only one respect. The vapor and condensate from the boiling liquid are passed through a **fractionation column** (Figure 6.1) before they reach the distillation head. This column contains a packing such as metal turnings or glass beads. As vapor rises through this column, it condenses on the packing and revaporizes continuously. Each revaporization of condensed liquid is equivalent to another simple distillation. Each of these "distillations" leads to a distillate successively richer in the lower-boiling component. Substantial enrichment of the vapor in the lower-boiling component occurs by the time the vapor reaches the head.

With a good fractionation column and proper operation, compounds with boiling points only a few degrees apart may be separated successfully. (Of course, an azeotropic mixture can *not* be separated into its components by fractional distillation, because an azeotrope is a mixture of constant composition with a constant boiling point).

6.1 Efficiency of the Fractionation Column

How well a fractionation column can separate a pair of compounds is called its **efficiency.** The efficiency of a column depends on both its length and its packing. In general, the greater the length, the greater its efficiency is. For columns of the same length, an increase in the surface or an increase in the heat conductivity of the packing results in an increase in efficiency.

Another factor affecting the efficiency of a distillation apparatus is the **reflux ratio.** This term is the ratio of the amount of material in the distillation head that undergoes *reflux* (condensing and flowing back onto the column)

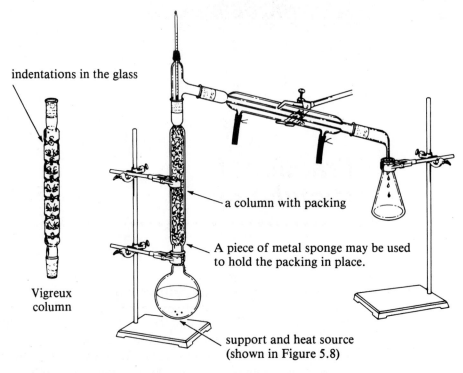

indentations in the glass

a column with packing

A piece of metal sponge may be used
to hold the packing in place.

Vigreux
column

support and heat source
(shown in Figure 5.8)

FIGURE 6.1   Apparatus for a fractional distillation.

to the amount of material that is removed through the sidearm as distillate.

$$\text{reflux ratio} = \frac{\text{weight of condensate at the top of the column that is returned to the column}}{\text{weight of the material at the top of the column that is removed as distillate}}$$

For example, a reflux ratio of 15 means that for every gram of distillate, 15 g
of the material drains back into the column. A higher reflux ratio results in
greater efficiency.

With commercial fractional distillation apparatuses, but generally not
with student apparatuses, the reflux ratio can be controlled. One type of
control device is a solenoid-operated glass valve that regulates the amount of
vapor passing into the condenser.

The efficiency of a column is reported in **theoretical plates,** where one
theoretical plate is equivalent to one simple distillation. A distillation assembly with an efficiency of two theoretical plates is therefore equivalent to two
simple distillations.

A distillation apparatus with one theoretical plate (a simple distillation
head) separates compounds with a difference in boiling points of about 100°
or more, but it does not give a good separation of compounds whose boiling
points are closer together. At the other extreme, a column of 100 theoretical
plates can separate a pair of compounds boiling as close as 2° apart. Most

TABLE 6.1   Number of theoretical plates (TP) needed to separate a binary mixture

Number of TP	Approx. bp difference (°C)
1	100
5	35
10	20
50	4
100	2

TABLE 6.2   Characteristics of some fractionation columns[a, b]

Type of column	Holdup (mL)	Theoretical plates (TP)	Height of each TP (HETP) (cm)	Separable bp difference (°C)
Vigreux	1.5	3	8	50
glass helices	5	6	4	30
metal sponge	9	6	4	30
spinning band[c]	0.2	11–61	0.4–2	3–20

[a] All values are approximate because of individual column characteristics, types of compounds to be separated, operational differences, etc.
[b] Values are based on columns 10 mm in diameter and 25 cm long.
[c] A very efficient type of column used in research laboratories. The wide range of values arises from the different models of these columns.

laboratory fractionation columns vary from 2 to about 15 theoretical plates. For example, a 25-cm column packed with glass beads or metal turnings has an efficiency of approximately 6 theoretical plates and, at best, can separate a binary mixture with a boiling-point difference of 30°–40°. Table 6.1 shows the number of theoretical plates needed to separate binary mixtures according to their boiling-point differences.

The number of theoretical plates is only one consideration in the choice of a column. Some fractionation columns have a greater *holdup* than others. This term refers to the quantity of condensed vapor that remains on the packing when the distillation is stopped. If a very small amount of liquid is to be distilled, a column with a minimum holdup, even though it is less efficient, might be the column of choice. One such column, which contains no packing, is the **Vigreux column** (see Figure 6.1). Table 6.2 summarizes the characteristics of some columns.

• • • • • • • • • •
6.2   Steps in Fractional Distillation

The following instructions for carrying out a fractional distillation assume that you will pack your own fractionation column. If a commercial distillation column is available, your instructor will provide directions for its use.

1) Packing the fractionation column. The technique for packing a fractionation column depends on the packing material. Metal turnings or sponges are best pulled into the column with a wire hook. If the packing is glass beads, glass helices, or small metal turnings, first place a piece of metal sponge at the bottom of the column to support the packing (see Figure 6.1); then pour or drop in the pieces of packing. Regardless of the type of packing used, it should be loosely, but uniformly, packed. "Holes" in the packing will decrease efficiency, while spots of very tight packing may plug the column or cause it to flood (see below).

2) Setting up the apparatus. Assemble the apparatus shown in Figure 6.1, with the fractionation column clamped in a vertical position. View the column from more than one direction to make sure it is vertical. When distilling high-boiling compounds, insulate the column with glass cloth, dry rags, or a double layer of loosely wrapped aluminum foil. Whenever practical, however, leave the column uncovered so that you can observe the behavior of the liquid–vapor mixture in the column.

Clamp the distillation flask ($\frac{1}{2}$ to $\frac{2}{3}$ full, containing boiling chips, and with its joint lightly greased) to the fractionation column. Clamp the receiving flask in position; then insert the thermometer into the distillation head.

• •

SAFETY NOTE Before heating, check that all joints are snug, that fresh boiling chips have been added, and that the system is open to the atmosphere at the receiver.

• •

3) The fractional distillation. Heat the distillation flask *slowly*. When the solution boils, you will observe the ring of condensate rising up the fractionation column. If heating is too rapid and the condensate is pushed up too rapidly, equilibration between liquid and vapor will not occur and separation of the components will not be satisfactory.

If you heat the distillation flask too strongly before the column has been warmed by hot vapors and condensate, the column may *flood*, or show an excessive amount of liquid in one or more portions of the packing. Flooding is due to lack of equilibration between condensate and vapor and is more likely to occur if the packing has not been inserted uniformly. Ideally, the packing should appear wet throughout, but no portion of it should be clogged with liquid.

Flooding can be stopped by lowering the heat source. As the boiling of the liquid diminishes, the excess liquid in the column flows back into the distillation flask. At this time, resume heating, but more slowly than before. If flooding recurs, insulate the column as described in Step (2) so that the vapors will have less tendency to condense. If the flooding is due to an incorrectly packed column, cool the apparatus, repack the column, and begin again.

4) *Collecting the fractions.* In a fractional distillation, read the boiling points and collect the fractions just as in a simple distillation. It is always better to collect a large number of small fractions than a few large ones. Small fractions of the same composition can always be combined, but a fraction that contains too many components must be redistilled.

Suggested Readings

Carney, T. P. *Laboratory Fractional Distillation.* New York: Macmillan, 1949.

Buck, A. C. "Efficiency of Fractional Distillation Columns." *J. Chem. Ed.* **1944,** 21(10), 475.

Linstead, R. P., Elvidge, J. A., and Whalley, M. *A Course in Modern Techniques of Organic Chemistry.* London: Butterworths, 1955.

Problems

6.1 Why is a fractionation column packed with glass helices more efficient than a Vigreux column of the same length and same diameter?

6.2 Which of the following circumstances might contribute to column flooding and why?
(a) "Holes" in column packing
(b) Packing too tight
(c) Heating too rapidly
(d) Column too cold

6.3 Explain why flooding in the fractionation column can lead to a poor separation of distilling components.

6.4 Refer to Figure 5.11 (page 107). In each of the distillations recorded in the figure, which fractions could be combined to yield reasonably pure compounds?

6.5 A chemist has a small amount of a compound (bp 65°) that must be fractionally distilled. Yet the chemist does not want to lose any of the compound to holdup on the column. What can the chemist do?

6.6 Referring to Tables 6.1 and 6.2, determine the approximate height of Vigreux columns needed to separate binary mixtures of compounds with the following boiling-point differences.
(a) 50°   (b) 25°   (c) 5°

Vacuum Distillation

When the pressure over a liquid is reduced, the liquid boils at a lowered temperature. (Refer to Figure 5.1, page 95, the boiling-point–pressure diagram for acetone.) As its name implies, **vacuum distillation** (simple or fractional) is distillation under reduced pressure. Because the reduction in pressure lowers the boiling point, vacuum distillation is used for distilling high-boiling or heat-sensitive compounds.

7.1 Boiling Point and Pressure

At pressures near atmospheric pressure, a drop in pressure of 10 mm Hg generally lowers the boiling point of a substance by about 0.5°. At the low pressures usually used in vacuum distillation, halving the pressure reduces the boiling point about 10°. For example, a compound with a boiling point of 100° at 20 mm Hg would boil at about 90° at 10 mm Hg.

Figure 7.1 shows a diagram, called a *nomograph*, for estimating boiling points at various pressures. To use the nomograph, follow these steps.

1) Assume that the boiling point of a compound at one particular pressure is known. Find the reported boiling point on scale A and the pressure (P_1) on scale C. Connect these two points with a transparent ruler.

2) Read the boiling point at atmospheric pressure (760 mm Hg) at the intercept on scale B.

3) To find the boiling point at a different pressure, P_2, connect the normal atmospheric boiling point on scale B to P_2 on scale C. Read the new boiling point from scale A.

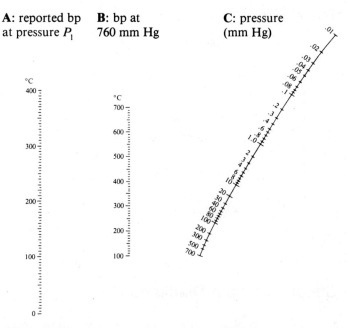

A: reported bp   **B:** bp at   **C:** pressure
at pressure P_1 760 mm Hg   (mm Hg)

FIGURE 7.1 A nomograph for estimating boiling points at different pressures.

EXAMPLE A compound boils at 80° at 1.0 mm Hg. What is its boiling point at 20 mm Hg?

1) Connect 80° on scale A to 1.0 mm on scale C.
2) Read the atmospheric boiling point on scale B; it is 250°.

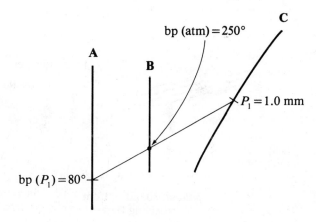

3) Connect 250° on scale B to 20 mm on scale C. Read the boiling point at 20 mm Hg from scale A. In our example, the estimated boiling point on scale A is 130°.

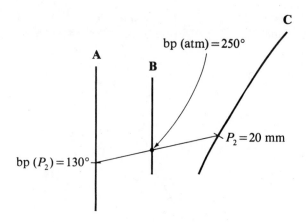

bp (atm) = 250°

A

B

C

P₂ = 20 mm

$P_2 = 20$ mm

bp (P_2) = 130°

7.2   Apparatus for Simple Vacuum Distillation

Figure 7.2 shows a typical student vacuum distillation apparatus. If the apparatus used in your laboratory is substantially different, your instructor will tell you how to assemble it.

Except for the adapter, a vacuum distillation apparatus is very similar to that used for atmospheric distillation. The principal difference is that, in vacuum distillation, the system must be airtight and under vacuum. Also, the

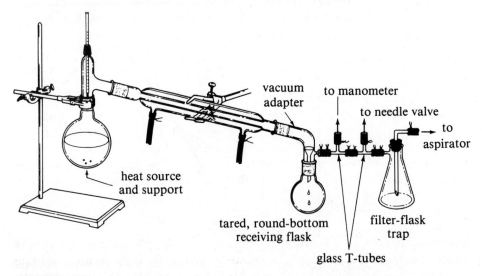

FIGURE 7.2   A simple vacuum distillation assembly. Each flask and the condenser are clamped in place (clamps not shown). All connecting heavy-walled rubber tubing is wired down.

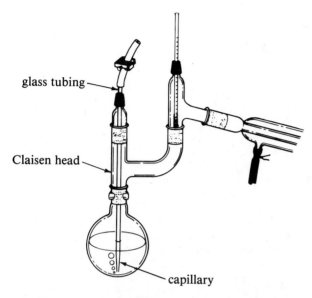

glass tubing

Claisen head

capillary

FIGURE 7.3   A distillation flask equipped with a very
fine capillary tube. When the apparatus is evacuated, a
thin stream of air bubbles is pulled through the
capillary to prevent bumping. For a fractional vacuum
distillation, a fractionation column can be inserted
between the Claisen head and the distillation head.

glassware must be able to withstand the vacuum; for example, large, thin-
walled Erlenmeyer flasks cannot be used. Under vacuum, most solvents boil
well below room temperature; therefore, any solvent in the material to be
distilled must be removed prior to distillation.

Use vacuum grease to seal all glass-to-glass connections. Stopcock
grease flows too easily under heat to hold a vacuum. Use only heavy-walled
vacuum tubing. Tighten all rubber-tubing connections (including a rubber
thermometer adapter) by wrapping a wire around the rubber and twisting it
with pliers. Assemble the apparatus completely and test for air leaks by
applying the vacuum before placing any liquid in the distillation flask.

Because ordinary boiling chips do not function well in a vacuum
distillation, special boiling chips with smaller pores must be used to prevent
bumping. Alternatively, a *very* fine capillary may be used to introduce a thin
stream of air or N_2 bubbles to the boiling liquid (see Figure 7.3). Do not pass
air through oxygen-sensitive compounds.

Vacuum adapter and fraction collectors   If a simple vacuum adapter is
used, the fore-run cannot be separated from the main fraction without
stopping the distillation and disassembling the apparatus. Therefore, if more
than one fraction is to be collected, a vacuum fraction collector should be
used. Two types of vacuum fraction collectors are shown in Figure 7.4.

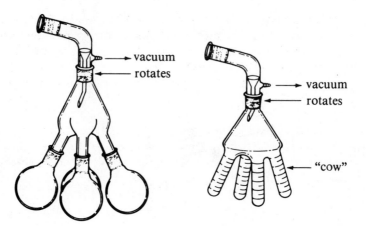

FIGURE 7.4 Vacuum adapters and receivers that allow the collection of more than one fraction during the course of a distillation.

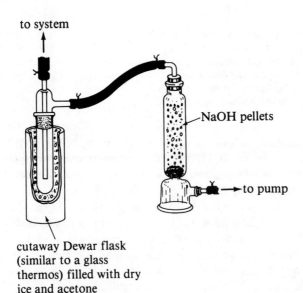

FIGURE 7.5 A dry-ice–acetone cold trap and a sodium hydroxide trap used to protect a mechanical vacuum pump. A dry-ice–acetone cold trap can be cooled to temperatures as low as $-80°$.

Vacuum sources   A vacuum can be obtained by using either a mechanical pump or a water aspirator. Mechanical pumps must be used for distillations at pressures lower than 20 mm Hg. If a mechanical pump is used, it must be protected by traps, such as the ones shown in Figure 7.5, to prevent vapors of distillate from passing into it. Your instructor can show you how to use the pump. For pressures of 20 mm Hg and above, a water aspirator is usually satisfactory.

One principal disadvantage of an aspirator results from fluctuating water pressure. If the water pressure drops, the vacuum in the distillation apparatus can suck water from the aspirator into the receiving flask. To ensure that such an event does not occur, position a water trap between the aspirator and the distillation apparatus (see Figure 7.2). This trap is similar to the trap for vacuum filtration, except that a one-hole stopper is used.

Pressure regulation and measurement   Place a needle valve (or other pressure regulation device) and a manometer (pressure-measuring device) between the vacuum source and the vacuum adapter. (A typical assembly using two T-tubes is shown in Figure 7.2.) The needle valve allows you to adjust the pressure to the desired level by bleeding air into the system. If a needle valve is not available, use the base of a Fischer or Meker burner. Connect the vacuum tubing to the gas inlet and adjust the vacuum with the valve at the base of the burner normally used to adjust the height of the flame. Pressure control with a needle valve or burner is satisfactory in distillations conducted at a pressure greater than 20 mm Hg.

Closed-end manometers, as shown in Figure 7.6, are suitable for measuring pressures above 20–25 mm Hg. In either of these manometers, the difference in the levels of mercury, in mm, is equal to the pressure in the system. For distillations at pressures lower than 20 mm, another pressure-

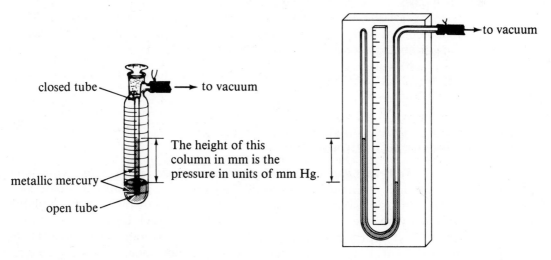

FIGURE 7.6   Typical closed-tube manometers.

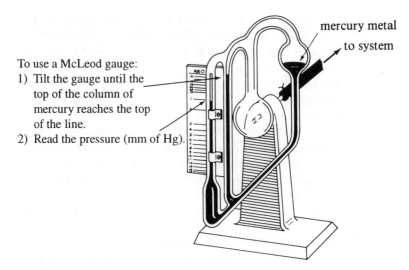

To use a McLeod gauge:
1) Tilt the gauge until the top of the column of mercury reaches the top of the line.
2) Read the pressure (mm of Hg).

mercury metal

to system

FIGURE 7.7 A tilting McLeod gauge, used for measuring very low pressures (0.01–10 mm Hg). The gauge does not automatically record changes in pressure but must be swiveled each time a pressure reading is made.

measuring device, such as a "tilting McLeod gauge," must be used (see Figure 7.7). The technique for using a McLeod gauge is outlined in Figure 7.7.

SAFETY NOTE   Whenever vacuum is applied to a closed system, there is the danger of an implosion. *Safety glasses must be worn at all times!* Before setting up the vacuum distillation apparatus, check all glassware for stars or cracks. (Only round-bottom flasks, pressure flasks, or Erlenmeyer flasks of 50-mL size or smaller should ever be used in a vacuum distillation.)

7.3   Steps in Vacuum Distillation

1) Test the vacuum with apparatus empty.   Tare a receiving flask and assemble a simple vacuum distillation apparatus (Figure 7.2) as described in the discussion. With an empty distillation flask, test the seals by applying vacuum with the needle valve closed. You should be able to obtain a pressure of approximately 20 mm Hg using an aspirator or a much lower pressure with a mechanical pump. If you cannot obtain this low pressure, use the pinch clamp to isolate various parts of the system to find the leak(s).

After checking the empty system, break the vacuum by opening the needle valve *slowly* while watching the mercury in the manometer rise. (If the vacuum is broken very rapidly, the rising mercury can have sufficient momentum to break the glass in the manometer.)

2) Add the mixture to be distilled.   Using a funnel with a stem, pour the material to be distilled (cool and solvent-free) into the distillation flask and add a few carbon or Micro-Porous® boiling chips if you are not using a capillary. Apply a full vacuum with the needle valve completely closed. You should be able to attain nearly the same low pressure as when the distillation flask was empty. If not, a new leak has developed; seal it before proceeding.

3) Set the pressure and distill.   Adjust the needle valve on the system so that the pressure is at the desired level; then let the system stand for a few minutes until the pressure is no longer changing. Heat the mixture to a boil with a heating mantle, and distill. (If the distillation is stopped and then resumed, add fresh boiling chips to the mixture.) At the end of the distillation, lower the heating mantle and allow the system to approach room temperature before breaking the vacuum, to avoid igniting the hot residue. Do not turn off the aspirator until the vacuum has been broken.

Suggested Reading

Vogel, A. I. *Practical Organic Chemistry*. 3rd ed. London: Longman Group Limited, 1956.

Problems

7.1   Near atmospheric pressure, a 10-mm drop in pressure usually lowers a boiling point by about 0.5°. Using this rule of thumb, predict the boiling points of the following compounds at the pressure indicated.
(a) Water at 745 mm Hg
(b) Dichloromethane (bp 41° at 760 mm) at 737 mm
(c) Benzene (bp 80° at 760 mm) at 765 mm

7.2   At low pressures such as used in vacuum distillation, reducing the pressure by half reduces a boiling point by about 10°. If a compound boils at 180° at 10 mm Hg, what would be its approximate boiling point at the following pressures?
(a) 15 mm Hg   (b) 2.0 mm Hg

7.3   Using the nomograph in Figure 7.1, approximate the boiling points of the following substances at atmospheric pressure (760 mm Hg).
(a) Rotenone, bp 210° (0.5 mm Hg)
(b) 1,6-Hexanediol, bp 132° (9 mm Hg)

7.4   Approximate the boiling points of the compounds in Problem 7.3 at 25 mm Hg.

7.5   Explain why a vacuum distillation apparatus should be checked for leaks before the material to be distilled is placed in the flask.

Steam
Distillation

. . . .

Steam distillation is the distillation of a mixture of water (steam) and an organic compound or a mixture of organic compounds. The organic compound must be insoluble (immiscible) in water for the steam distillation to be successful. Immiscible mixtures such as water and an organic compound do not behave like solutions. The components of an immiscible mixture boil at *lower temperatures than the boiling points of any of the components*. Thus, a mixture of high-boiling organic compounds and water can be distilled at a temperature less than 100°, the boiling point of water. Natural products, such as flavorings and perfumes found in leaves and flowers, are sometimes separated from their sources by steam distillation. This technique can sometimes replace vacuum distillation and other separation techniques in the organic laboratory.

. . . .
8.1 Characteristics of Steam Distillation

The principle that gives rise to steam distillation is that the total vapor pressure of a mixture of immiscible liquids is equal to the *sum of the vapor pressures of the pure individual components*. The total vapor pressure of the mixture thus equals atmospheric pressure (and the mixture boils) at a lower temperature than the boiling point of any one of the components singly. For a steam distillation to be successful, however, the component to be isolated must have a vapor pressure higher than those of the other components in the mixture (such as polymeric material or inorganic salts).

This boiling behavior of an immiscible mixture is different from that of miscible liquids, where the total vapor pressure is the sum of the *partial* vapor pressures of the components. Therefore, with immiscible mixtures, the mole fraction of the components in the mixture is not important.

For immiscible liquids:

$$P_{total} = P_A^0 + P_B^0 \qquad \text{— } \textit{vapor pressures of pure components}$$

For miscible liquids:

$$P_{total} = \underbrace{X_A P_A^0}_{} + \underbrace{X_B P_B^0}_{} \quad \text{— } \textit{partial vapor pressures}$$

In a distillation of two immiscible liquids, the number of moles of each component in the vapor (and thus in the distillate) is proportional to the vapor pressure of the pure component.

For the steam distillate:

$$\frac{\text{moles of A}}{\text{moles of B}} = \frac{P_A^0}{P_B^0}$$

8.2 Calculation of the Amount of Water Needed to Distill an Organic Compound

The previous equations allow us to calculate the actual amount of water needed to steam-distill a particular compound. Substituting the definition of moles and rearranging the equation, we see that the weight ratio of the two components is dependent only on the molecular weights and vapor pressures.

$$\frac{wt_A/MW_A}{wt_B/MW_B} = \frac{P_A^0}{P_B^0}$$

$$\frac{wt_A}{wt_B} = \frac{P_A^0 \times MW_A}{P_B^0 \times MW_B}$$

In order to calculate how much water is needed, we need only the vapor pressure of water at the distillation temperature, and vapor pressure of the compound to be distilled at that temperature, and their molecular weights. The vapor pressure of water at various temperatures is well known. A few of these values are listed in Table 8.1. (*The Handbook of Chemistry and Physics* contains a more extensive table.)

In general, the vapor pressure of the compound will not be known, but it can be estimated using the vapor pressure of water and the total atmospheric pressure.

TABLE 8.1   Vapor pressure of water at various temperatures

Temperature (°C)	Vapor pressure (mm Hg)
60	149
70	234
80	355
85	434
90	526
95	634
100	760

$$P_B^0 = P_{\text{total}} - P_A^0$$

with an annotation: *vapor pressure of water at the distillation T* pointing to P_A^0.

If the structure of the organic compound is known, then its molecular weight can be calculated. The molecular weight of water, of course, is well known.

EXAMPLE   At 760 mm Hg, bromobenzene (C_6H_5Br, bp 155°) steam-distills at approximately 95°. Calculate the approximate amount of water needed to steam-distill 20 g of bromobenzene.

$$\frac{\text{wt}_{H_2O}}{\text{wt}_{C_6H_5Br}} = \frac{P_{H_2O}^0 \times MW_{H_2O}}{P_{C_6H_5Br}^0 \times MW_{C_6H_5Br}}$$

$$= \frac{634 \times 18}{(760 - 634) \times 157}$$

$$= \frac{0.58 \text{ g } H_2O}{\text{g } C_6H_5Br}$$

Thus, for 20 g of C_6H_5Br, we would need to distill 20×0.58 g, or about 12 g, of water.   ●

8.3   Apparatus for Steam Distillation

The simplest way to perform a steam distillation is to add water to the material to be distilled and distill in the ordinary manner. However, if a large

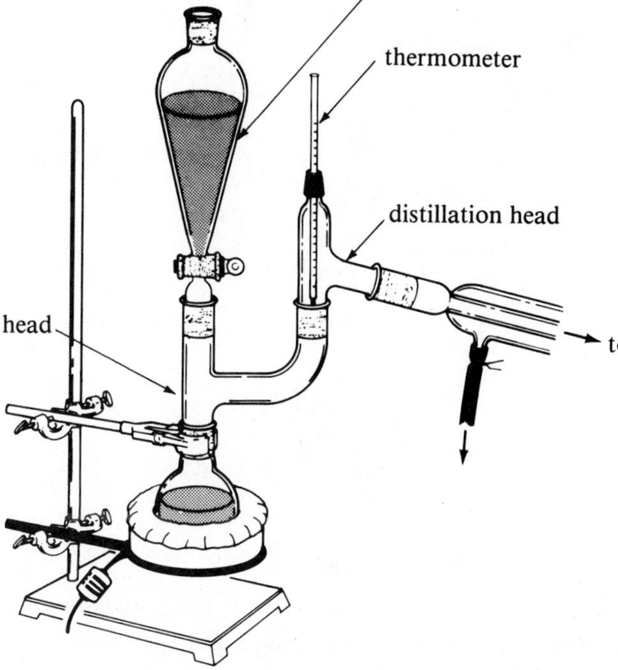

dropping funnel containing water
(The water is added to the distillation
flask periodically to maintain its level.)

thermometer

distillation head

Claisen head

to receiver

FIGURE 8.1 A steam distillation apparatus that allows water to be added
to the distillation flask during the distillation.

volume of water is needed, the water may boil away before the steam
distillation is completed. Figure 8.1 shows a setup that allows the water in
the distillation flask to be replenished.

Live steam can also be used in a steam distillation. Figure 8.2 shows a
distillation setup that allows the use of a steam line. (CAUTION: The steam
inlet is hot.) One advantage of using live steam is that the mixture need not
always be heated to boiling, even though some heating is necessary to
prevent excessive condensation of the steam in the flask. The trap shown in
Figure 8.2 also helps prevent excessive water from entering the flask by
trapping condensed steam.

The quantity of distillate collected in a steam distillation depends on the
vapor pressure of the organic component. A compound with a high vapor
pressure will steam-distill with a small amount of water. If the vapor
pressure of the organic material to be isolated is low, a greater amount of
distillate must be collected. The steam distillation is complete when only
pure water distills. This may be determined by noting when the distillate is
clear and no longer composed of two phases.

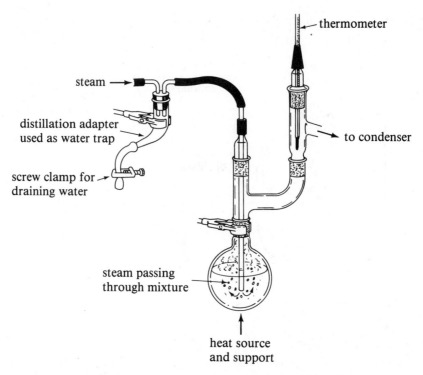

FIGURE 8.2 A steam distillation apparatus for using live steam.

Suggested Readings

Morton, A. A. *Laboratory Techniques in Organic Chemistry*. New York: McGraw-Hill, 1938, Chapter 6.

Vogel, A. I. *Practical Organic Chemistry*. 3rd ed. London: Longman Group Limited, 1956.

Problems

8.1 A mixture of immiscible liquids (both water-insoluble) is subjected to steam distillation. At 90°, the vapor pressure of pure water is 526 mm Hg. If the vapor pressure of compound A is 127 mm Hg and that of B is 246 mm Hg at 90°:
(a) What is the total vapor pressure of the mixture at 90°?
(b) Would this mixture boil at a temperature above or below 90°?
(c) What would be the effect on the vapor pressure and boiling temperature by doubling the amount of water used?

8.2 Suggest a reason why benzene steam distills at the rate of 1 g/0.1 g H_2O, but nitrobenzene distills at the rate of 1 g/4 g H_2O.

8.3 (a) At 99°, the vapor pressure of water is 733 mm Hg. At standard atmospheric pressure, what is the vapor pressure of a compound being steam-distilled at this temperature?

(b) If the compound has a molecular weight of 180, how much water is required to distill 1.0 g of the compound?

8.4 At 99.6°, water has a vapor pressure of 750 mm Hg and quinoline (immiscible with water) has a vapor pressure of 10 mm Hg. What weight of water must be distilled for each gram of quinoline in a steam distillation at 760 mm Hg?

quinoline

Sublimation

Sublimation is a process whereby a solid is purified by vaporizing and condensing it without its going through an intermediate liquid state.

Solid compounds that evaporate (that is, pass directly from the solid phase to the gaseous phase) are rather rare; solid CO_2 (dry ice) is a familiar example of such a compound. Even though both solids and liquids have vapor pressures at any given temperature, most solids have very low vapor pressures. In order for a solid to evaporate, it must have an unusually high vapor pressure compared with those of other solids. For a solid compound to exhibit such a high vapor pressure, it must have relatively weak intermolecular attractions. One factor that contributes to weak intermolecular attractions is the shape of the molecules. Many compounds that evaporate readily contain molecules that are roughly spherical or cylindrical—shapes that do not lend themselves to strong intermolecular attractions. Table 9.1 lists some solids that can be sublimated in the laboratory.

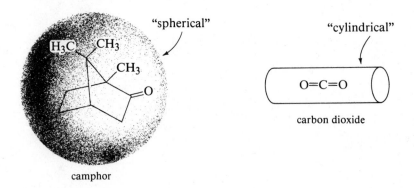

camphor

Sublimation can be used to purify some solids, just as distillation can be used to purify liquids. In sublimation, nonvolatile solid impurities remain

TABLE 9.1 Vapor pressures of some solids at their melting points

Compound	mp (°C)	Vapor pressure (mm Hg) at the melting point[a]
hexachloroethane	186	780
camphor	179	370
iodine	114	90
p-dichlorobenzene	53	8.5
naphthalene	80	7
benzoic acid	122	6

[a]The melting point of a compound is the limiting temperature for its sublimation.

behind when the sample evaporates, and condensation of the vapor yields the pure solid compound. Sublimation has the advantages of being reasonably fast and clean because no solvent is used. Unfortunately, most solid compounds have vapor pressures too low for purification in this fashion. Also, sublimation is successful only if the impurities have much lower vapor pressures than that of the substance being purified. It would be practical, for example, to purify technical-grade iodine, which is contaminated with inorganic salts, by sublimation. It would *not* be practical, however, to separate camphor from isoborneol (from which commercial camphor is synthesized) because *both* compounds readily sublime.

isoborneol

The vapor pressure of a solid increases with temperature, just as it does for liquids. Therefore, evaporation can be facilitated by heating the solid, but

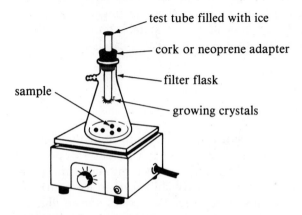

FIGURE 9.1 A sublimation apparatus using a filter flask, ice-filled test tube, and hot plate.

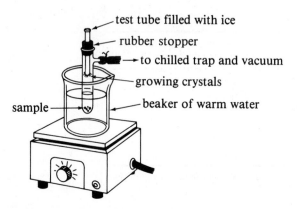

FIGURE 9.2 A vacuum sublimation apparatus using a large sidearm test tube.

not to its melting point. The rate of evaporation can also be increased by subliming the solid under a vacuum; however, a very efficient cooling surface must be used for the condensation so that the solid's vapor is not lost into the vacuum system.

A simple sublimation apparatus for a student laboratory is shown in Figure 9.1. Figure 9.2 shows an apparatus for carrying out a sublimation under vacuum.

9.1 Atmospheric Sublimation

The atmospheric sublimation apparatus, shown in Figure 9.1, can be used only for a solid with a relatively high vapor pressure. Place the sample in a filter flask equipped with a water-cooled cold finger or a test tube filled with ice. Warm the flask on a hot plate or in a water bath, taking care not to melt the solid. Crystal growth on the test tube (and on the cooler flask sides) soon occurs. Periodically, cool the apparatus, remove the cold finger or test tube carefully, and scrape the sublimed crystals into a suitable tared container. Determine the melting point of the sample with a sealed capillary.

9.2 Vacuum Sublimation

For a vacuum sublimation, choose a sidearm test tube to hold the sample, and insert a smaller test tube fitted into a large-holed rubber stopper (see Figure 9.2). Place ice in the inner tube, and connect the sidearm, using heavy-walled rubber tubing, to a trap (preferably chilled in an ice bath) and then to an aspirator or other vacuum source. (See Section 1.2 for an illustration of the trap and a discussion of using the aspirator.)

Suggested Readings

Linstead, R. P., Elvidge, J. A., and Whalley, M. *A Course in Modern Techniques of Organic Chemistry*. London: Butterworths, 1955.

Robertson, G. R. "Sublimation." *J. Chem. Ed.* **1932,** *9(10),* 1713.

Kern, D. M. "The Heat of Sublimation of Carbon." *J. Chem. Ed.* **1956,** *33(6),* 272.

Problems

9.1 Which of the following compounds could be subjected to sublimation at atmospheric pressure?
(a) Compound A: vapor pressure at its melting point = 770 mm Hg
(b) Compound B: vapor pressure at its melting point = 400 mm Hg
(c) Compound C: vapor pressure at its melting point = 10 mm Hg

9.2 In the preceding problem, which compounds could be sublimated with vacuum?

9.3 Which of the following compounds would be likely to evaporate readily? Explain your answer.

(a) $CH_3(CH_2)_{14}CH_3$
hexadecane

(b)
fenchone
(in oil of fennel)

(c)
pentacene

TECHNIQUE 10

Microscale
Procedures

. . . .

Microscale procedures involve reactions that use just a few milligrams of chemicals. Success on a microscale depends on the use of techniques and apparatus that minimize handling. Every time a solid or liquid is transferred from one container to another, there are opportunities for loss; therefore, transfers must be kept to a minimum.

. . . .
10.1   Reaction Vessels

A microscale reaction is carried out in a small (5- or 10-mL) pear-shaped flask or in a heavy-walled, conical reaction vial (see Figure 10.1). When a pear-shaped flask is used, it should be supported in a small cork ring or placed into a small beaker. Alternatively, the flask should be clamped to a support rod.

Reaction vials come in several sizes and have a threaded cap. The cap has an opening through which the bottom of a reflux condenser can be inserted. An O-ring in the cap is used to make a firm seal.

. . . .
10.2   Weighing and Transferring
A. Solids

For microscale reactions, a sensitive, top-loading balance should be used to weigh the solid to the nearest milligram. When a solid is to be weighed into an empty flask, first tare the empty flask; then add a small amount of the

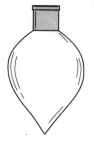 

pear-shaped flask   conical reaction vessel

FIGURE 10.1 Microscale reaction vessels. (From *Theory and Practice in the Organic Lab* by Landgrebe. Copyright © 1993 by Wadsworth, Inc. Reprinted by permission.)

FIGURE 10.2 Stainless steel microspatulas. (From *Theory and Practice in the Organic Lab* by Landgrebe. Copyright © 1993 by Wadsworth, Inc. Reprinted by permission.)

solid with a microspatula and check the new weight. A stainless steel microspatula (Figure 10.2), rather than a standard-sized spatula, should be used to minimize loss while transferring the solid.

To save time when the solid is the limiting reagent, weigh the solid to only a few milligrams of the specified weight. Adjustments in the amounts of the other reagents will not be necessary provided they are in excess.

B. Liquids

Automatic-delivery pipets (Figure 10.3) are used for delivery of fixed volumes of a liquid, a selection of several fixed volumes of a liquid, or variable volumes of a liquid. Many of these pipets can be read to at least 10 μL (10^{-3} mL). They draw the liquid into a polyethylene tip, which can eventually be discarded. Some pipets are equipped with an automatic tip ejector so that you do not need to handle a contaminated tip.

The following procedure should lead to reproducible results provided that the liquid is nonviscous.

1) Depress the button at the top of the pipet until resistance is encountered (first stop).

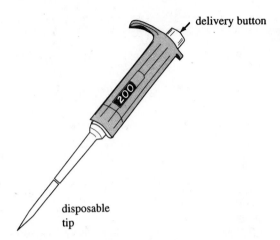

2) Dip the tip a few millimeters below the surface of the liquid and carefully release the button to draw up the liquid. Never let the button release with a snap. There should be no air bubbles in the tip; if there are, try again or switch to a new tip. After the liquid has been drawn into the tip, carefully wipe off any liquid on the outside of the tip with an absorbent tissue.

3) Put the tip of the pipet into the container into which delivery is desired and push the button slowly past the first stop. Touch the tip to the inside of the container to deliver the last drop.

Gas chromatography syringe If the liquid is not corrosive, a 10-μL gas chromatography syringe (Figure 10.4) can be used to deliver an accurate amount of liquid to a vessel. The metal plunger of the syringe prohibits the use of corrosive liquids.

Pasteur and filter pipets When accurate delivery is not needed, small amounts of liquids can be conveniently transferred by using a Pasteur pipet (Figure 10.5). For liquids with a higher vapor pressure (which tends to make them then squirt out of the pipet), a filter pipet (see Figure 10.5) gives better control. The filter pipet contains a very small tuft of cotton that has been pushed through the pipet into the tip with a piece of 20-gauge copper wire. (Straighten the wire by holding it firmly at both ends with pliers, snapping it, and then cutting off the ends.) Practice in selecting the correct amount of cotton is needed. For the best results, use cotton that has been rinsed with a small amount of methanol and hexane to remove contaminants and impurities and then dried.

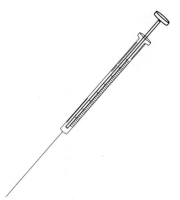

FIGURE 10.4 Gas chromatography syringe. (From *Theory and Practice in the Organic Lab* by Landgrebe. Copyright © 1993 by Wadsworth, Inc. Reprinted by permission.)

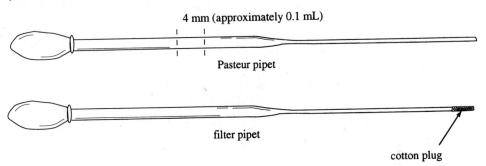

4 mm (approximately 0.1 mL)

Pasteur pipet

filter pipet

cotton plug

FIGURE 10.5 Pasteur pipet and filter pipet. (From *Theory and Practice in the Organic Lab* by Landgrebe. Copyright © 1993 by Wadsworth, Inc. Reprinted by permission.)

10.3 Measuring Physical Constants

A. Melting Points

The procedure for measuring a melting point on a microscale is the same as that described in Technique 2.

B. Boiling Points

One of the best methods for determining a boiling point is the microscale technique patterned after the method reported by A. Siwoloboff. The method requires only a few microliters of liquid and is carried out in a conventional melting-point capillary tube in a melting-point apparatus.

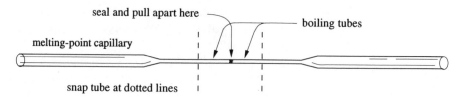

FIGURE 10.6 Boiling-tube preparation. (From *Theory and Practice in the Organic Lab* by Landgrebe. Copyright © 1993 by Wadsworth, Inc. Reprinted by permission.)

Prepare a sealed capillary about 1 cm in length that will fit into the melting-point capillary. To prepare the sealed capillary, soften the center sections of a melting-point capillary tube in a microburner flame (adjusted to show a blue cone). Remove the softened capillary from the flame and pull it at least several inches apart (see Figure 10.6). Place the center of the narrowed portion of this thin tube in the flame and pull it apart (into two pieces), thus sealing the ends. Snap each narrow tube about 1 cm from the sealed end to obtain two boiling tubes. Insert one of the boiling tubes, open end down, into a melting-point capillary and let it fall (or tap it) to the bottom.

Place the sample into the melting-point capillary with a 10-μL gas chromatography syringe or with a pipet made from a melting-point capillary that has been drawn out to a long, thin tip with a burner. If a gas chromatography syringe is used, the sample may be deposited as a slug of liquid, which will then have to be centrifuged to the bottom of the capillary. In this case, place the melting-point capillary in a centrifuge tube, open end up. Balance the centrifuge with another tube opposite it; spin the centrifuge up to speed and then let it come to rest.

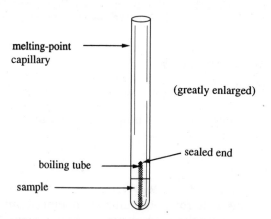

FIGURE 10.7 Boiling-point determination. (From *Theory and Practice in the Organic Lab* by Landgrebe. Copyright © 1993 by Wadsworth, Inc. Reprinted by permission.)

You should now have a sample tube that looks like the one in Figure 10.7. Place the melting-point capillary into a standard melting-point apparatus and begin to slowly heat the sample toward the expected boiling point. When you observe bubbles streaming from the open (bottom) end of the boiling tube within the melting-point capillary, turn off the heater and allow the sample tube to cool. At the moment when the bubbles stop forming and the liquid begins to be sucked back into the boiling tube, read the thermometer. The temperature should be within 1° of the correct boiling point if the sample is pure.

· · · · · · · · · · ·
10.4   Purification Procedures

A. Crystallization

The steps in microscale crystallization are completely parallel to the steps used in crystallization of a much larger sample (see Technique 1). The following procedure is satisfactory for quantities of 10–100 mg provided that the volume of hot solvent required to dissolve the crude solid does not exceed 1.5 mL. (See J. A. Landgrebe, *J. Chem. Ed.* **1988**, *65*, 460.) The method requires two Pasteur pipets, a piece of 20-gauge copper wire about 30 cm in length, a heat lamp, a micro three-finger clamp (or other comparable small clamp), a 20 × 150-mm sidearm test tube, and a small vial (0.5 dram).

The procedure for microscale crystallization follows.

1) Prepare a straight copper wire.   Straighten the wire by snapping it while gripping both ends with pliers. Trim any bent ends. Cut the wire into two pieces, one about 6 cm and the other about 24 cm. The shorter piece will be used as a stirring rod later in the procedure.

2) Prepare a filter pipet.   (See Section 10.2B.)

3) Prepare a recrystallization tube.   Push a cotton plug (about twice as large as the plug used for the filter pipet) so that it is positioned 1–2 cm into the narrowed tip of the pipet (see Figure 10.8). Then place the portion of the pipet toward the tip (about 2–3 cm below the cotton plug) into a flame of a microburner, and seal the tube by pulling it apart. This operation results in a very narrow tip that can be broken easily at the appropriate time. The cotton plug will filter out the insoluble impurities (see below).

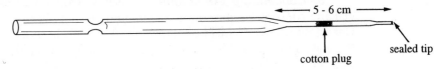

FIGURE 10.8   Preparation of a recrystallization tube. (From *Theory and Practice in the Organic Lab* by Landgrebe. Copyright © 1993 by Wadsworth, Inc. Reprinted by permission.)

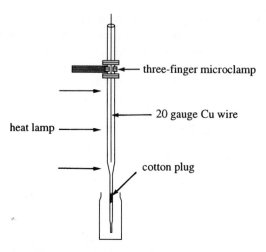

three-finger microclamp

20 gauge Cu wire

heat lamp

cotton plug

FIGURE 10.9   Recrystallization tube being used.
(From *Theory and Practice in the Organic Lab* by
Landgrebe. Copyright © 1993 by Wadsworth,
Inc. Reprinted by permission.)

4) *Add the solid to be recrystallized.*   Place the crude solid into the weighed recrystallization tube and weigh the tube again to determine how much solvent will be needed (see Technique 1). Clamp the tube near the top in a vertical position (see Figure 10.9) so that the bottom end is about 2 cm above a cork ring or wood block on the desk. Position a weighed, uncapped vial under the crystallization tube so that the tip of the tube extends about 1 cm into the vial.

5) *Dissolve the solid.*   Add an appropriate amount of solvent into the recrystallization tube. Stir the suspension of solvent and solid with the short wire, and use a heat lamp to carefully bring the solvent in the tube to a boil. Mount the heat lamp horizontally about 6–8 cm from the recrystallization tube and vial, and control the heat intensity of the lamp with a variable transformer. When the solid has dissolved, move the vial to the side to snap off the tip of the recrystallization tube, allowing the warm solution to drain into the vial. The contents of the recrystallization tube will drain through the cotton plug, filtering the solution. Continue heating the solvent just below its boiling point during this filtration step. If the filtration is too slow, apply a little pressure with a pipet bulb to the top of the recrystallization tube.

6) *Concentrate the solution.*   Insert a wooden boiling stick into the vial and concentrate the solution by boiling. If there is any possibility that the vial might tip over while boiling, clamp it or place it in a small beaker. Allow the concentrated solution to cool slowly, and finally, chill it with ice.

7) *Remove the mother liquor.*   When crystallization is complete, place a bulb on a filter pipet and draw up the mother liquor in the vial. It may take a minute or two to draw up all the liquid. Keep the vial cold during this

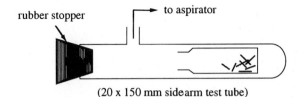

(20 x 150 mm sidearm test tube)

FIGURE 10.10 Sidearm test tube used as a vacuum drying oven. (From *Theory and Practice in the Organic Lab* by Landgrebe. Copyright © 1993 by Wadsworth, Inc. Reprinted by permission.)

operation. Add a small amount of cold, fresh solvent to wash the crystals, and remove the wash solvent as before.

8) Dry the crystals. Place the vial with the crystals in it into a 20×150-mm sidearm test tube attached to an aspirator hose, as shown in Figure 10.10. You may also wish to place a trap between the aspirator and the sidearm test tube (see Technique 1). Apply vacuum to dry the crystals. Careful application of heat from the lamp will dry the sample quickly. Care must be taken not to heat the sample above its melting point. A final weighing should be carried out to determine the yield of the pure product.

This procedure involves no transfer of the hot solution prior to filtration and no transfer of the filtered crystalline product to a storage vial. With practice, a recrystallization can be completed in 20–30 minutes, exclusive of drying time.

Microscale crystallization with a Craig tube One method for crystallizing 10–100-mg quantities employs a Craig tube (see Figure 10.11) and a centrifuge. The lower (smaller-diameter) part of a Craig tube is available with volumes of 1, 2, and 3 mL. The inner glass bulb fits snugly into the other tube and seats against a ground-glass surface where the tube narrows (see Figure 10.11). Less breakable inner sections made of Teflon® are also available.

Steps in the use of the Craig tube follow:

1) Dissolve the solid. Place the sample in a small vial (0.5 dram) or test tube (10×75 mm), and dissolve it in somewhat more than the minimum amount of solvent at just below the boiling point. Heating can be accomplished with a sand bath or with a shielded heat lamp. Stir the mixture with a microspatula or boiling stick.

2) Transfer the solution to a Craig tube. Warm a filter pipet by drawing hot solvent into it or by heating it with a heat lamp. Transfer the hot solution from the vial (or test tube) to the lower portion of a Craig tube.

3) Crystallize the solid. Add a boiling stick and concentrate the solution to perhaps one-half to one-third of the original volume (or until crystalline material begins to form easily). Insert the upper bulb of the Craig tube, and allow the entire apparatus to cool slowly in a safe place. The tube

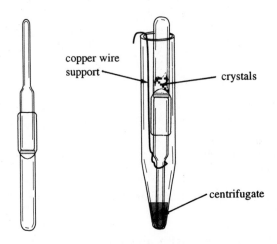

Craig tube   inverted Craig tube in a centrifuge tube

FIGURE 10.11   A Craig tube and a Craig tube in a centrifuge tube. (From *Theory and Practice in the Organic Lab* by Landgrebe. Copyright © 1993 by Wadsworth, Inc. Reprinted by permission.)

can be supported in a small Erlenmeyer flask or beaker. The use of a seed crystal may be necessary. Cooling in ice can improve the yield but should not be done until the tube reaches room temperature.

4) *Isolate the crystals.*   To isolate the crystals, place a centrifuge tube over the Craig tube and invert the whole assembly, as shown in Figure 10.11. Spin the assembly in a centrifuge to separate the liquid centrifugate (mother liquor) from the crystalline suspension. Disassemble the apparatus and isolate the crystals. Allow the crystals to air-dry, or place them in a suitable sidearm test tube for vacuum drying, as shown in Figure 10.10.

B. Extraction

When extractions are carried out with approximately 1-mL volumes of solvent, losses during separation of the layers can become significant regardless of the size of the separatory funnel. For that reason, it is usually best to carry out such an extraction by shaking the organic solvent and water layers in a small vial with a screw cap lined with polyethylene or Teflon®. (An alternative to using the liner is to stretch a thin film of polyethylene across the top of the vial before the cap is screwed into place.) If a reaction has been carried out in a small pear-shaped flask, the extraction can sometimes be accomplished by simply adding water and swirling.

After mixing, the layers are allowed to separate, and the lower layer is carefully removed by use of a Pasteur pipet (see Figure 10.12). Depending on

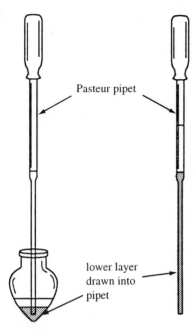

Pasteur pipet

lower layer
drawn into
pipet

FIGURE 10.12 Microscale extraction. (From *Theory and Practice in the Organic Lab* by Landgrebe. Copyright © 1993 by Wadsworth, Inc. Reprinted by permission.)

the solvent densities, the lower layer may be the organic or the aqueous layer. Plan to remove only the lower layer. Even with a steady hand, it is extremely difficult to quantitatively remove the upper layer with a pipet.

When the total volume of both liquid layers is less than the capacity of the pipet, it is best to draw up both layers into the pipet for separation. When both layers are in the pipet, the demarcation between the layers can be more easily seen, and the lower layer can be delivered into a small vial or flask for temporary storage (see Figure 10.12).

If you encounter difficulty in trying to find the line of demarcation between the aqueous and organic layers, add a small amount of copper sulfate to the aqueous layer to give it a distinct blue color.

Figures 10.13 and 10.14 illustrate the transfer steps required in microscale extraction with two successive portions of an organic solvent. Figure 10.13 shows the steps needed for a lighter-than-water solvent, and Figure 10.14 shows the steps needed for a heavier-than-water solvent.

Once the organic layers have been combined, drying is accomplished by adding 100–200 mg of a conventional drying agent (see Technique 4), such as anhydrous sodium sulfate or magnesium sulfate. Mix the suspension thoroughly and allow it to stand for 5–10 minutes, and then remove the organic layer with the aid of a filter pipet.

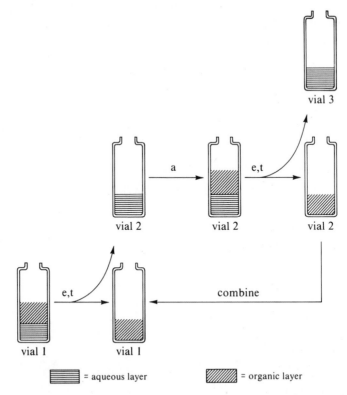

= aqueous layer   = organic layer

(a = add fresh organic solvent; e = extract (cap, shake, let settle); t = transfer lower layer)

FIGURE 10.13 Microscale extraction procedure using an organic solvent that is lighter than water. (From *Theory and Practice in the Organic Lab* by Landgrebe. Copyright © 1993 by Wadsworth, Inc. Reprinted by permission.)

C. Distillation

Simple distillation   Hickman distillation apparatuses, or stills, for separating a volatile solvent from a nonvolatile or very slightly volatile solute on a microscale are shown in Figure 10.15. In a Hickman distillation apparatus, the volatile liquid distills from the pot and is collected on a small trough or ring above the pot. Usually, the size of the receiving trough will accommodate from 0.1 to 1 mL. The glass surface directly above the trough is kept cool by wrapping it with a narrow strip of cloth on the outside glass surface and saturating the cloth with an aqueous solution of ethanol or acetone.

If you are to obtain a reasonable boiling point, the bulb of the thermometer must be entirely below the inner lip of the trough. A Hickman distillation apparatus can be conveniently heated by using a vertically mounted, shielded heat lamp. Other methods of heating include the use of a sand bath.

When the distillation is finished, the distillate is removed from the trough with a Pasteur pipet.

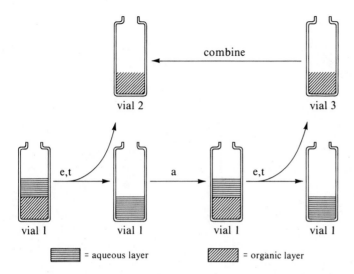

(a = add fresh organic solvent; e = extract (cap, shake, let settle); t = transfer lower layer)

FIGURE 10.14 Microscale extraction procedure using an organic solvent that is heaver than water. (From *Theory and Practice in the Organic Lab* by Landgrebe. Copyright © 1993 by Wadsworth, Inc. Reprinted by permission.)

Fractional distillation   Distillation apparatuses that can be used on a slightly larger scale (1–50 mL) are shown in Figure 10.16. The apparatus is sometimes called a short-path still because the vapor path between the distillation flask and the receiver is short. Because of the compact design, little distillate will be lost on the inner surfaces of the distillation head or condenser. A sidearm below the condenser can be connected to a vacuum source for distillation at a reduced pressure. The receiver can be a simple flask or a rotating collector

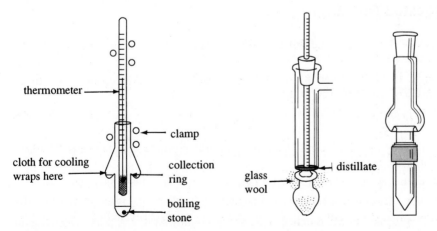

FIGURE 10.15 Hickman distillation apparatus for microscale distillations. (From *Theory and Practice in the Organic Lab* by Landgrebe. Copyright © 1993 by Wadsworth, Inc. Reprinted by permission.)

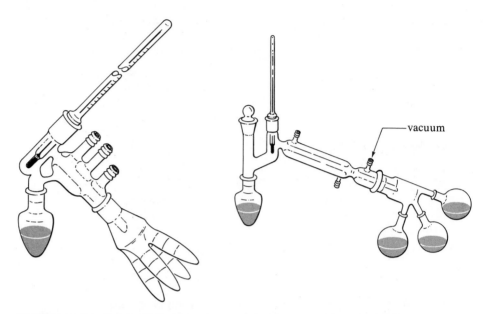

FIGURE 10.16 Small distillation units with rotating collectors for fractional distillation. (From *Theory and Practice in the Organic Lab* by Landgrebe. Copyright © 1993 by Wadsworth, Inc. Reprinted by permission.)

known as a *pig* or *cow*. The latter type of collector has a special advantage with distillations at reduced pressure: a simple rotation changes the receiver without disrupting the vacuum (and the vapor–liquid equilibrium), as would occur if the receiver flasks had to be changed.

Suggested Readings

Mayo, D. W., Pike, R. M., and Butcher, S. S. *Microscale Organic Laboratory*. 2nd ed. New York: Wiley, 1989.

Lawler, R. G., and Parker, K. A. "Efficient Distillation in Microscale Laboratory," *J. Chem. Ed.* **1986**, *63(11)*, 1012.

Volter, E. J., and Cook, D. "A Microscale Sample Transfer Method." *J. Chem. Ed.* **1988**, *65(6)*, 538.

Craig, R. E. R. "Rapid, Efficient Determination of Recrystallization Solvents." *J. Chem. Ed.* **1989**, *66(1)*, 88.

Ruekberg, B. "Robust Micro-Lab Sand Baths." *J. Chem. Ed.* **1989**, *66(1)*, 89.

Winston, A. "Design of a Microscale Sublimator." *J. Chem. Ed.* **1990**, *67(2)*, 162.

Mayo, D. W., Pike, R. M., Butcher, S. S., and Meredith, M. L. "A Microscale Organic Laboratory. III. Simple Procedure for Carrying Out Ultra-Micro Boiling Point Determinations." *J. Chem. Ed.* **1985**, *62(12)*, 1114.

Landgrebe, J. A. "Microscale Recrystallizations with a Disposable Pipet." *J. Chem. Ed.* **1988**, *65*(5), 460.

Schwartz, M. H. "Microscale Distillation—Calculations and Comparisons." *J. Chem. Ed.* **1991**, *69*(4), A127.

Siwoloboff, A. "Ueber die Siedepunkbegtimmung kleiner Mengen Flüssigkeiten." *Ber.* **1886**, *19*, 795a.

Problems

10.1   What two types of reaction vessels are used in microscale reactions?

10.2   Use the following microscale reaction to answer the questions.

acetanilide	110 μL	500 μL	*p*-nitroacetanilide
1.48 mmole	1.65 mmole		

(a) What is the limiting reagent?
(b) If you weigh out 180 mg of acetanilide, do you need to (increase/ decrease/ leave as shown) the volume of nitric acid?
(c) Can you use a gas chromatography syringe to deliver the two acids into the reaction vessel?

10.3   Why is an automatic-delivery pipet generally used for the delivery of nonviscous liquids?

10.4   How should you dispose of the disposable tip of an automatic-delivery pipet?

10.5   In preparing a filter pipet, why don't you wash the cotton used for the plug with soap and water rather than with methanol and hexane?

10.6   In the preparation of a boiling tube for a microboiling-point determination, what would happen if the tube was not sealed at one end?

10.7   What is the purpose of each of the following in a recrystallization tube?
(a) The cotton plug
(b) The sealed tip

10.8   Why is it undesirable to separate small volumes of immiscible liquids with a separatory funnel?

10.9   In the extraction chart shown in Figure 10.13, explain why vial 2 and not vial 3 is combined with vial 1.

10.10 Can a Hickman distillation apparatus be used for the separation of liquids with similar boiling points?

Refractive
Index

A refractive index is a physical property, like a boiling point, that can be used as one piece of evidence in determining the identity and purity of a liquid compound. **Refraction** is the bending of a ray of light as it passes obliquely from one medium to another of different density. Refraction arises from the fact that light travels more slowly through a more dense substance. In organic chemistry, we are interested in the refraction of light as it passes through a liquid layer. Refraction is useful because the degree of refraction depends upon the structure of the compound.

The **refractive index** (n) of a particular substance is defined as the ratio of the speed of light in a vacuum or in air (the values are very close) to the speed of light in the substance.

$$\text{refractive index, } n = \frac{\text{speed of light in air}}{\text{speed of light in the substance}}$$

The refractive index is measured with an instrument called a **refractometer,** which determines the angle of refraction of light between the liquid and a prism. A typical refractometer is shown in Figure 11.1.

Different wavelengths of light are refracted different amounts. This is the reason that sunlight can be split into the visible spectrum (a rainbow) by droplets of water. When the refractive index is used as a physical constant, only one wavelength of light is used, usually the sodium D line, at 589.3 nm. This single wavelength can be obtained from either a sodium lamp or an ordinary white light with a prism system.

Besides being dependent on wavelength, the refractive index is temperature-dependent. For this reason, the temperature is always specified when a refractive index is reported. A typical refractive index for a compound, such as for benzene or ethanol, is reported as in the following

prism section
(obscured in this drawing);
opens for sample insertion

focusable eyepiece

movable lamp

disperson correction
control

read display button

temperature
display button

on/off mode selector

adjustment control

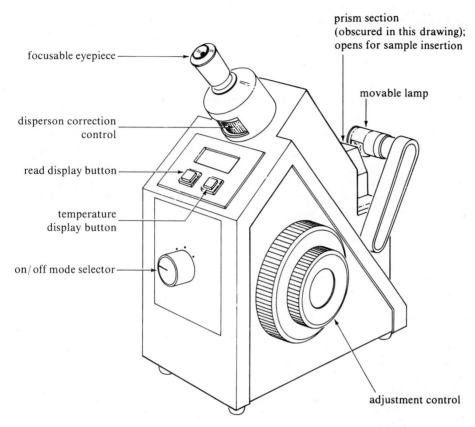

FIGURE 11.1 A typical refractometer (Courtesy of Leica Inc., Buffalo, New York. Reprinted by permission.)

examples:

benzene, n_D^{20} 1.5011   ethanol, n_D^{20} 1.3611

where D refers to the wavelength (the sodium D line) and the superscript number is °C

The refractive index is a very sensitive physical property. Unless a compound is extremely pure, it is almost impossible to duplicate a literature value exactly. For example, a sample of benzene from a fractional distillation might exhibit n_D^{20} 1.4990 instead of the reported value of n_D^{20} 1.5011. However, the closer the observed refractive index is to the reported value, the more pure the compound is likely to be.

In terms of structure, the refractive index is a function of the polarizability of the atoms and groups within molecules. A more polarizable molecule exhibits a higher refractive index. Thus, we find that alkyl iodides (with large polarizable iodo- substituents) and most aromatic compounds (with polarizable pi systems) have high refractive indexes, generally greater than 1.5000.

By contrast, most aliphatic compounds exhibit refractive indexes between 1.3500 and 1.5000.

11.1 Correcting for Temperature Differences

The refractive index varies inversely with temperature. An increase in temperature causes a liquid to become less dense and almost always causes a decrease in refractive index. It has been determined experimentally that the average temperature correction factor for a wide variety of compounds is 0.00045 unit for each degree Celsius. The following examples show how to adjust a refractive index to a different temperature.

EXAMPLE 1
(Adjustment to a higher temperature)

To adjust n_D^{18} 1.4370 to n_D^{20}, proceed as shown.

Calculate factor:

$$2°C \times 0.00045/°C = 0.00090$$

Subtract from measured value:

$$n_D^{20} = 1.4370 - 0.0009$$
$$= 1.4361$$ ●

EXAMPLE 2
(Adjustment to a lower temperature)

To adjust n_D^{23} 1.5066 to n_D^{20}, proceed as shown.

Calculate factor:

$$3°C \times 0.00045/°C = 0.00135 \quad (\text{round to } 0.0014)$$

Add to measured value:

$$n_D^{20} = 1.5066 + 0.0014$$
$$= 1.5080$$ ●

11.2 Steps in Using a Refractometer

Your instructor will give you specific instructions on the use of the refractometer in your laboratory. The following general steps, however, apply to most refractometers.

1) Prepare the prism section. Open the prism section, which can be done mechanically. If the prism surfaces are not absolutely clean, clean them with a few drops of ethanol and a lens-cleaning tissue. (CAUTION: The prism surfaces are very easily scratched. Do not touch these surfaces with anything hard, such as an eye dropper, the end of a glass rod, or a metal spatula. Clean them gently by dabbing with the tissue.)

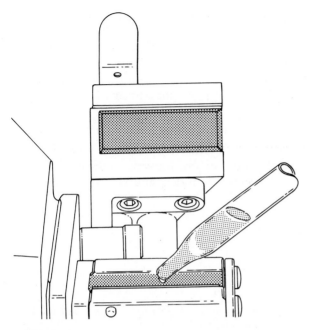

FIGURE 11.2 An open prism section with the sample being applied to the lower surface of the prism. (Courtesy of Leica Inc., Buffalo, New York. Reprinted by permission.)

2) *Place the liquid on the prism surface.* Place 1–3 drops of the liquid to be analyzed on the lower prism surface; then close the prism section. The prism surfaces fit closely together; when the section is closed, the liquid is forced to spread out as a thin film. Or close the prism section and then apply a drop or two of liquid to the edge of the gap between the prism surfaces. A thin film will form between the surfaces by capillary action. Figure 11.2 shows an open prism section with the sample being applied to the lower prism surface.

3) *Position the light.* Position the lamp to shine directly into the glass prism of the instrument.

4) *Adjust the instrument.* Look through the eyepiece and use the appropriate knobs to focus the cross hairs and to bring a light area and a dark area into view. The appearance of the field is illustrated in Figure 11.3. If the dividing line between the light and dark areas is fringed with red or blue, the prism system is not directing the sodium D line exactly through the sample. The prisms may be adjusted by a dispersion correction knob. The field is in proper adjustment when there is a sharp dividing line, with no color, between the light and dark areas.

the cross hairs

the light and
dark areas

properly adjusted for reading
the refractive index

FIGURE 11.3   The field of view through the eyepiece of a refractometer.

5) *Center the dividing line.* To determine the refractive index, move the dividing line between the light and dark areas *exactly* to the center of the cross hairs, read the value for the refractive index, and record the temperature.

6) *Clean the instrument.* (See Step 1).

Suggested Readings

Weissberger, A., ed. *Technique of Organic Chemistry. I. Physical Methods*. 3rd ed. New York: Interscience Publishers, 1959, Chapter 15.

Wendland, R. T. "A System of Characterization of Pure Hydrocarbons. Refractometric Analysis as a Key to the Evaluation of Structure." *J. Chem. Ed.* **1946**, *23*, 3.

Maley, L. E. "Refractometers," *J. Chem. Ed.* **1968**, *45(6)*, A467.

Hartkopf, A., Schroeder, R. R., and Meyers, C. H. "Quantitative Analysis of Xylene Mixtures by Refractometry," *J. Chem. Ed.*, **1974**, *51(6)*, 405.

George, S. W., and Campbell, J. A. "Refractive Indices of Some Carbon Compounds as a Function of Temperature." *J. Chem. Ed.* **1967**, *44(7)*, 393.

Problems

11.1   Correct the following refractive indexes to 20°C.
(a) n_D^{22} 1.4398   (b) n_D^{30} 1.4702   (c) n_D^{16} 1.3962   (d) n_D^{18} 1.4022

11.2   Compound A has a refractive index of 1.4577 at 20°, while compound B (miscible with compound A) has a refractive index of 1.5000 at the same temperature.
(a) Experimentally, how could you determine if the refractive indexes of mixtures of A and B follow a linear relationship to the mole percent compositions of the mixtures?
(b) If the relationship is linear, what is the composition (in mole percent) of a mixture that has a refractive index of 1.4678 at 20°?

11.3   When measuring a refractive index, what should you do if (a) no dividing line between light and dark areas can be observed through the eyepiece, or (b) the dividing line can be seen, but it is fringed with color?

Gas–Liquid Chromatography

Chromatography is a general term that refers to a number of related techniques used for analyzing, identifying, or separating mixtures of compounds. The use of chromatographic techniques is not limited to organic chemistry —these techniques find wide use in a variety of scientific areas. For example, gas–liquid chromatography is used in criminology laboratories for blood alcohol tests; thin-layer chromatography and gas–liquid chromatography are used in environmental and biology laboratories; and all types of chromatography are used in medical laboratories for both research and routine analyses. A relatively new tool, high-performance liquid chromatography (HPLC), combines many of the best features of two chromatographic techniques: gas–liquid chromatography and column chromatography.

All chromatographic techniques have one principle in common: a liquid or gaseous solution of the sample, called the **moving phase,** is passed (moved) through an adsorbent, called the **stationary phase.** The different compounds in the sample move through the adsorbent at different rates because of physical differences (such as vapor pressure) and because of different interactions (adsorptivities, solubilities, and so on) with the stationary phase. Thus, the individual compounds in a sample become separated from one another as they pass through the adsorbent and can be either collected or detected, depending on the chromatographic technique and the quantity of sample used.

In this technique, we will discuss gas–liquid chromatography. In the techniques that follow, we will discuss column chromatography and two related techniques: flash chromatography and HPLC (Technique 13) and thin-layer chromatography (Technique 14).

151

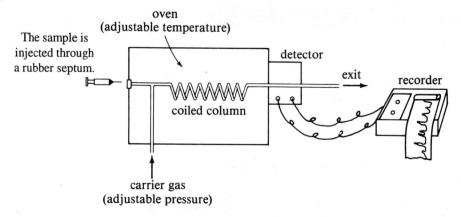

FIGURE 12.1 The basic components of the gas chromatograph. The vaporized sample is carried through the column by an inert carrier gas. The passage of organic compounds is sensed by the detector, and this information is graphed by the recorder.

12.1   Gas–Liquid Chromatography

Gas–liquid chromatography (GLC) is known by a variety of other names: *gas chromatography (GC)*, *gas–liquid phase chromatography* or *gas–liquid partition chromatography (GLPC)*, and *vapor phase chromatography*. All these terms refer to the same technique, which is an instrumental method of analyzing the components of a mixture. The instrument, called a **gas chromatograph,** is diagramed in Figure 12.1.

About 1–10 μL* of liquid sample or solution is injected with a small hypodermic syringe into the gas chromatograph, through a rubber septum on the heated injection port of the instrument. The sample is vaporized and carried through a heated column by an inert carrier gas (usually helium or nitrogen). The adsorbent in the column is a high-boiling liquid suspended on a solid inert carrier. Because of differing interactions with the adsorbent and differing vapor pressures, the components of the sample move through the column at different rates. At the end of the column, each component passes through a detector, which is connected to a recorder. The recorder produces a tracing that shows *when* each component of a mixture passes the detector and also indicates the *approximate relative quantity* of each component.

GLC is commonly used for checking the purity of volatile-liquid samples, such as distillation fractions; checking on the identity of a substance by comparison of its GLC with that of a known; and analyzing a mixture for

---

*1 microliter (μL) = 1×10^{-6} L, or 1×10^{-3} mL.

the presence or absence of a known compound (such as alcohol in blood). In addition, the relative quantities of the components of a mixture can be determined with about 5% accuracy.

In most GLC units, small quantities of sample are used and no attempt is made to collect the material that has been chromatographed. However, with *preparative gas chromatographs*, the separation and collection of samples are possible. Also, special techniques allow the effluents from a gas chromatograph to be further analyzed—for example, by a mass spectrometer.

Typical liquid phases used in GLC columns:

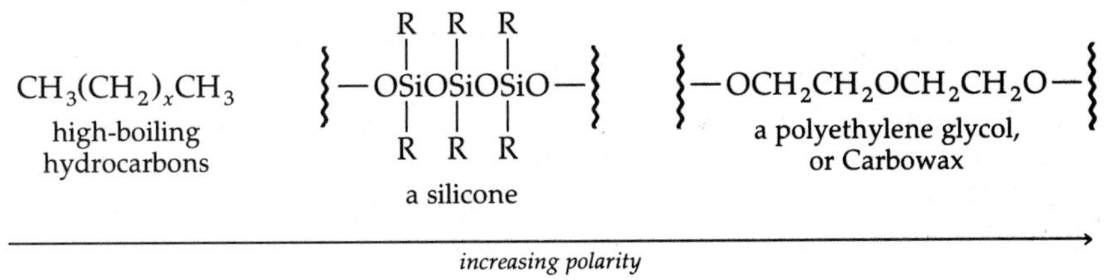

$$CH_3(CH_2)_xCH_3 \qquad \left\{-OSiOSiOSiO-\right\} \qquad \left\{-OCH_2CH_2OCH_2CH_2O-\right\}$$

high-boiling hydrocarbons

a silicone

a polyethylene glycol, or Carbowax

increasing polarity →

The liquid phase of a GLC column does not truly adsorb the gaseous sample. Instead, the sample dissolves in the liquid phase. The dissolved compound is in equilibrium with its vapor, a process that depends primarily on the compound's vapor pressure. (Henry's law: the solubility of a gas in a liquid is directly proportional to the pressure of that gas above the liquid.)

As the vapor of a sample is carried through the column by the carrier gas, it continuously dissolves and revaporizes. A nonpolar compound with a high vapor pressure moves along a nonpolar column at a fairly rapid rate. A less volatile compound moves more slowly. Thus, the compounds tend to

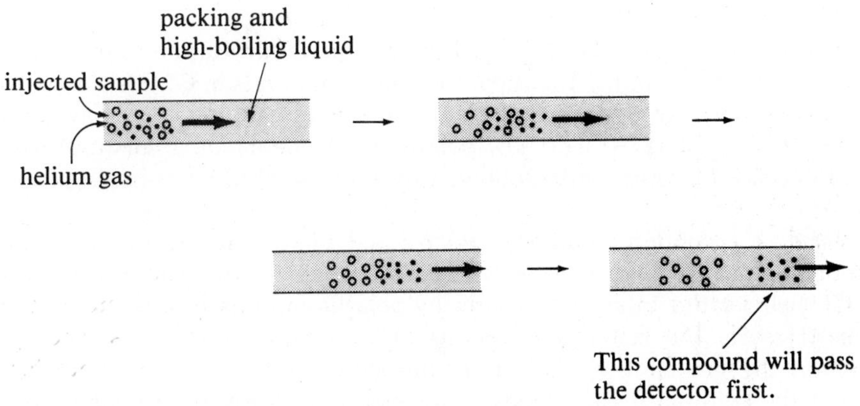

packing and
high-boiling liquid

injected sample

helium gas

This compound will pass the detector first.

FIGURE 12.2 A mixture of compounds becomes separated into individual components as it passes through a GLC column.

concentrate in bands as they move through the column, as illustrated in Figure 12.2.

· · · · · · · · · · ·
12.2   The Gas Chromatograph
A. The Column

The column of a typical gas chromatograph is constructed of metal tubing, which is coiled so that a long column (3–10 feet) will fit into the small volume of the heated oven. The column contains an inert but loosely packed solid (such as crushed firebrick or diatomaceous earth) that is coated with the liquid phase. The inert packing provides spacing and allows the carrier gas to flow through the column without the need for excessive pressure. The liquid phase is the active "component" in the column (see below). A number of interchangeable columns can be purchased for a gas chromatograph so that the appropriate packing liquid and column length can be chosen for a particular separation.

In the instrument, the column is located in an insulated oven with adjustable temperature controls. The upper useful limit of temperature is determined by the vapor pressure of the liquid phase in the column. The usual operating temperature is 100°–200°, but higher temperatures may be used with some types of columns. The operating temperature chosen depends on the boiling points of the compounds in the sample to be chromatographed.

A typical liquid phase used in a GLC column is Silicone Oil DC-550, which is useful for a wide variety of compounds. For polar compounds, a more polar liquid phase, such as a Carbowax, may be used. Columns containing less polar liquids, such as mineral oil, are occasionally used for the GLC analysis of nonpolar compounds.

In columns that have a high-boiling nonpolar hydrocarbon as the liquid phase (a nonpolar column), the separation of compounds in GLC is similar to their separation in distillation, because both processes depend primarily on relative vapor pressures. The low-boiling components of a mixture pass through the GLC column first, followed by successively higher-boiling components.

When the column contains a polar-liquid phase, the separation of the components in the mixture depends upon (1) their relative vapor pressures and (2) their relative interactions with the polar-liquid phase. The greater the vapor pressure, the faster the compound will pass through the column. However, the greater the interaction, the slower the compound will pass through the column. Because these two effects are unrelated (a compound can be low-boiling and polar, for example or high-boiling and nonpolar), it is often very difficult to predict the order in which compounds will elute from the column.

B. The Detector

The detector at the end of the column signals when a compound is being eluted from the column. A number of different types of detectors have been invented, but only two types are in common use: **thermal conductivity (hot-wire) detectors** and **flame ionization detectors.**

In a thermal conductivity detector, an electric current is passed through a wire located directly in the flow of gaseous effluent from the column. The electrical resistance of a hot wire varies with the heat conductivity of a gas passing over it. Helium, with a high heat conductivity, absorbs heat from the wire and keeps it relatively cool. Organic compounds have lower heat conductivities. Therefore, when the vapors of an organic compound pass over the wire, the wire becomes hotter; consequently, its resistance changes. The hot wire is actually one arm (the sample arm) of a *Wheatstone bridge* circuit (see Figure 12.3). The other arm (the reference arm) is a similar hot wire with pure helium (no sample) passing over it. When no sample is passing through the sample arm, the bridge is in balance and the recorder

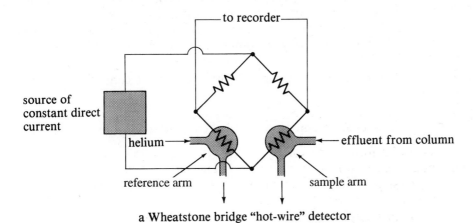

a Wheatstone bridge "hot-wire" detector

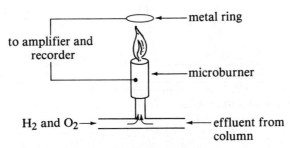

a flame ionization detector

FIGURE 12.3 Two common types of GLC detection devices.

pen rides on the baseline. The passage of a gaseous organic compound from the column through the sample arm of the detector changes the resistance of the sample arm. Thus, the bridge becomes unbalanced, and an electric current passes to the recorder.

In a flame ionization detector (Figure 12.3), part of the gaseous effluent from the column is mixed with a separate gas stream of hydrogen and oxygen, and this gaseous mixture is burned. When organic compounds are present in the effluent from the column, they are oxidized and converted to ions by the high temperature of the flame. The ions then pass through a metal ring, creating an electrical potential difference between the ring and the barrel of the burner. The small potential difference is amplified and then sent to the recorder.

C. The Chromatogram

The *recorder* receives the electrical information from the detector and produces a graph (the **chromatogram**) of the components passing through the detector. Figure 12.4 is a representation of a typical gas chromatogram.

When injecting a sample, indicate the start of the chromatogram by a small reference mark. Three ways to generate the reference mark follow.

1) Move the recorder pen slightly with the baseline control at the time of injection.

2) Using a separate pen or pencil, make a mark on the recorder paper at the time of injection.

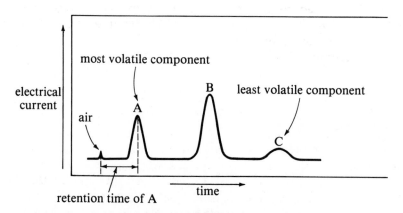

FIGURE 12.4 A gas chromatogram using a nonpolar column. The air peak signals the time of injection. The retention time of a compound is measured as the distance between the air peak or other reference mark and the top of the compound peak, converted to minutes.

3) Inject a small air bubble along with the sample. This air will pass through the column rapidly; when it passes the detector, the recorder pen automatically traces a small peak. A small air peak is shown on the chromatogram in Figure 12.4.

These three types of reference marks may not be positioned identically and thus may not be quantitatively comparable. So that you can compare different chromatograms, use one type of reference mark consistently. Then, on each chromatogram, note the type of reference mark used.

Retention time   The time it takes for a particular compound to pass through the column after its injection and reach the detector is called the compound's **retention time.** The retention time is a function of the vapor pressure of the compound, the rate of gas flow, the temperature, the liquid phase, and the length and diameter of the column.

The retention time is measured from the reference mark to the top of the compound peak. Retention times may be reported in minutes (which are the correct units) or in units of distance, such as centimeters (because the recorder paper moves at a constant speed). For example, if the paper moves at 1.27 cm/minute, a measured retention time of 5.2 cm equals 4.1 minutes. If the retention time is reported in units of distance, the speed of the recorder must also be reported.

$$\text{retention time} = \frac{\text{number of cm from injection to peak}}{\text{speed of the paper (cm/minute)}}$$

Under carefully controlled conditions, retention times can be duplicated. If accuracy is desired, an inert compound such as a hydrocarbon can be added to the sample and *relative retention times* measured from the internal standard's peak instead of from the operator's mark or an air peak. With this technique, relative retention times can be reproduced with an accuracy of about 1%.

12.3 Steps in a GLC Analysis

1) Preparing the instrument.   A gas chromatograph cannot be turned on and used immediately. The injection port, the detector, and the oven must be allowed to come to thermal equilibrium. For this reason, one or more of the heat sources (depending on the instrument) and a slow flow of carrier gas should be started well in advance of instrument use. Your instructor will probably prepare the instrument before class.

Each time your make a run on the instrument, make sure that the gas flow is adjusted properly, the temperature is at the proper level, the auxiliary equipment (detector, recorder) is switched on, and the recorder pen is tracing a baseline. Your instructor will demonstrate these manipulations.

• •

SAFETY NOTE   Cylinders of compressed gas must always be chained to the wall or laboratory bench so that they cannot be knocked over. The valves should not be handled until their use has been demonstrated.

• •

2) Injecting the sample.   Use a microhypodermic syringe to inject a sample through the silicone rubber septum into the vaporizing chamber. The amount to inject depends on the capacity and sensitivity of the instrument; a few microliters is usually sufficient. (CAUTION: Microsyringes are delicate and expensive. The metal plunger and needle are easily bent. Dirt in the glass barrel of the syringe will cause the plunger to stick; if the plunger is forced, the barrel will split. Your instructor will demonstrate the proper handling and cleaning procedures for microsyringes.)

Inject the sample quickly, so that it vaporizes all at once. With a larger microsyringe (100 μL), hold the plunger before and after the injection. (The carrier gas in some instruments is under enough pressure that it can cause the plunger to fly out.) Some instruments are inconsistent in showing an automatic injection signal; therefore, mark the position of the pen on the recorder paper with a soft pencil or felt pen when you make an injection.

If a *solution* (instead of an undiluted sample mixture) is injected, a large solvent peak will be observed at the start of the chromatogram. Do not worry if the recorder pen runs to the top of the recorder paper and whines; the pen will return to the baseline when the solvent has all passed the detector.

3) Obtaining the chromatogram.   Once an injection is made, an operator has little to do but watch the tracing of the pen. Do not make another injection until the previous one has passed through the column.

After the chromatogram has been recorded, stop the flow of paper from the recorder and cut or tear off your chromatogram. Mark it with your name, the date, the name of the sample, the sample size, the column type (silicone, Carbowax, and so on), the temperature of the column, the flow rate of the carrier gas, the speed of the recorder, and any comments you feel necessary.

4) Calibration of gas chromatograms.   There are two ways to calibrate a gas chromatogram to identify components. The first is to run a chromatogram of a pure sample of known composition immediately before or after the chromatogram of the unknown sample. The conditions (temperature, gas flow, and so on) should not be changed during the calibration run and actual run. The second calibration method assumes that a sample has already been subjected to GLC analysis, so that the appearance of the gas chromatogram is known. A drop of the known is mixed with a few drops of the sample, and the chromatogram of this mixture is then compared with the original chromatogram. If the addition of the known has increased the size of one of the existing peaks, the compound giving rise to the peak may be identical with the known. If the addition of the known compound to the mixture results in a new peak, then that known compound is not present in the sample. While such identification techniques are useful in the laboratory, they are not considered an absolute proof of structure. Corroborating evidence is needed.

12.4   Problems Encountered in GLC

Most of the problems encountered in GLC analysis are easily correctable. *Poor resolution*, in which peaks overlap each other, may be the result of too large a sample, too high a gas flow, or too high a temperature. In some cases, a different type of column (for example, Carbowax instead of silicone oil) or a longer column with a narrower diameter may be necessary for good resolution.

A peak that runs off the top of the recorder paper is the result of too large a sample or too high a sensitivity setting. Most instruments have a sensitivity control, called an *attenuator*, which can be adjusted if necessary. Increasing the attenuator setting decreases the peak height. If the peak is too high, either reinject with a smaller volume or increase the attenuator and reinject using the same volume.

A peak that is very broad or unsymmetrical can result from the injection of too much material. A compound with an usually long retention time may also be broad and sometimes very flat. In this case, the gas flow should be rechecked. Increasing the temperature or switching to another type of column usually solves this problem.

Some problems in GLC work can be traced to a worn-out injection septum. The carrier gas and the sample can both leak out through holes or cracks in the septum. Leakage leads to small sample peaks (or none at all) and nonreproducible retention times.

If an injected sample gives rise to no recorder signal and if the instrument is otherwise operating normally, the sample may have been retained in the column. A retained sample can result in contamination of future analyses, a change in the character of the column, and even a plugged column. For these reasons, avoid injecting tars or polymers into a GLC column. (In fact, your instructor may ask you to analyze only *distilled* samples by GLC.) Notify your instructor if your sample does not pass through the column.

12.5   Uses of Gas Chromatograms
A. Qualitative GLC Analyses

As we have mentioned, GLC is commonly used for checking the purity of a volatile sample. Determining the actual identity of a sample component from a GLC analysis cannot be done directly. In a research laboratory, it may be necessary to isolate a compound, such as by preparative GLC, and to determine its structure by physical constants, spectral data, and chemical analysis. In routine laboratory analyses, when we already have a good idea of the components in a mixture, such a lengthy identification procedure is usually not necessary.

In some cases, the structure of a compound giving rise to a peak in a GLC chromatogram can be determined indirectly. If the compound is known to belong to a particular homologous series, it is possible to identify the compound by comparing the log of its retention time with those of three (or more) other members of the same homologous series. For example, suppose that the sample compound is a methyl ester of a carboxylic acid with a continuous, saturated chain of unknown length. We would select three known methyl esters of continuous-chain carboxylic acids and plot the logs of their retention times versus the number of carbons in their chains. The result is a straight line. The number of carbons in the chain of the unknown ester can then be determined from the graph and the log of its retention time.

Besides showing the minimum number of components in a sample and their retention times, a gas chromatogram obtained from a nonpolar column can show the approximate relative boiling points. If the instrument is calibrated with similar compounds of known boiling points, we can closely estimate the actual boiling points of the components in the sample.

B. Quantitative Analysis

The area under a peak in a gas chromatogram is directly proportional to (1) the amount of compound in the sample and (2) the detector response. Detector response varies with different types of compounds. If the variations in detector response are ignored, then we can calculate the compound's percent easily, but our accuracy is not great (around 10%). Accuracy can be

width (w) at $\frac{1}{2}$ peak height, in cm

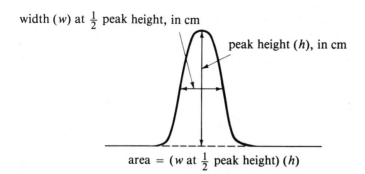

peak height (h), in cm

area $=$ (w at $\frac{1}{2}$ peak height) (h)

FIGURE 12.5 The area of a GLC peak is approximately equal to the width of the peak at its half height multiplied by its full height.

improved by calibrating the instrument by injecting known amounts of the compounds and calculating a correction factor for detector response.

Determining peak areas   Regardless of whether or not the chromatogram is to be corrected for detector response, peak areas—not peak heights—must be used for quantitative calculations. Peak heights depend upon retention time; peak areas do not. Many gas chromatographic recorders are equipped with mechanical or electronic **integrators** that automatically provide peak areas. If your recorder has one of these devices, your instructor will show you how it is used.

If a gas chromatograph does not have an integrator, the area can be approximated with an accuracy of about 3% by considering the peak to be a *triangle* (area $=\frac{1}{2}$ base $\times$ height). Instead of measuring the base of a GLC peak and dividing by 2, use the width of the peak at $\frac{1}{2}$ peak height, because it yields more reproducible values (see Figure 12.5). Determining areas by triangulation works well only if the peaks are *well-resolved* and *symmetrical*. (The Suggested Readings, p. 166 list sources that describe determination of the areas of overlapping peaks.)

Another technique for determining the relative areas under peaks, including unsymmetrical peaks, is to carefully cut out the peaks (or tracings of the peaks on plain, uniform bond paper) and weigh them. The weights of the peaks are proportional to the areas.

With either technique, the recorder paper should be set to a fast speed to maximize peak width and minimize errors.

Quantitative analysis without correcting for detector response   If a rough approximation of the percents of each component in a sample is satisfactory, then the area of each component is expressed as a percent of the total area of all components in the chromatogram. Because of its simplicity, this is the method most commonly used in quantitative GLC analysis.

EXAMPLE   A sample produces a chromatogram with three peaks, 8.0, 0.75, and 7.2 cm^2 (see Table 12.2). What is the percent of the three components in the mixture?

$$\frac{8.0 \text{ cm}^2}{(8.0 + 0.75 + 7.2)\text{cm}^2} \times 100 = \frac{8.0 \text{ cm}^2}{16 \text{ cm}^2} \times 100 = 50\%$$

$$\frac{0.75 \text{ cm}^2}{16 \text{ cm}^2} \times 100 = 5\%$$

$$\frac{7.2 \text{ cm}^2}{16 \text{ cm}^2} \times 100 = 45\% \qquad \bullet$$

Correcting areas with an internal standard   The weight ratios or weight percents of components in a mixture can be determined indirectly from a gas chromatogram by considering the *areas under the peaks* and *correction factors* that compensate for different detector responses. Although there are several ways to convert area ratios to corrected weight ratios, we will present only one: the **internal-standard technique.** A detector produces a signal that is directly proportional to the quantity of compound injected. Because different compounds exhibit different proportionality constants, it is necessary to obtain correction factors for quantitative determinations of mixture composition. The internal-standard technique utilizes weight and area ratios for these corrections.

In the internal-standard technique, a reference compound, or standard, is added to both a calibrating mixture and the mixture to be analyzed. Any compound can be used as a standard. However, for best results, the standard should be structurally similar to the components being analyzed and should be present in the mixture in approximately the same concentration. Also, the instrument should be adjusted so that the peak of the standard is well separated from the other peaks in the chromatogram.

The internal-standard technique involves three general steps. First, the calibrating mixture of *known weights* of the components to be analyzed and the standard is prepared; the chromatogram of this known mixture is then obtained. Next, the weights and peak areas for this known mixture are used to calculate a correction factor for each component, which can be applied to a chromatogram of a mixture of unknown percent composition. Finally, the chromatogram of the mixture of unknown percent composition (also containing the reference compound) is obtained, and the percent composition of this mixture is calculated.

Let us consider the procedure in more detail. Assume that we wish to analyze a mixture of compounds A, B, and C to determine its percent composition. The specific steps in the procedure follow.

For the known mixture:

1)   Add a weighed amount of standard to a mixture containing known weights of compounds A, B, and C.

2)   Obtain the gas chromatogram and determine the areas of the peaks.

3)   Calculate the weight ratio (component/standard) and the area ratio (component/standard) for each component.

for A:

$$\text{wt ratio (A/standard)} = \frac{\text{wt of A}}{\text{wt of standard}}$$

$$\text{area ratio (A/standard)} = \frac{\text{area of A's peak}}{\text{area of standard peak}}$$

Similar ratios are calculated for B and C.

4)   For each component, calculate the weight/area correction factor.

$$\text{correction factor, } F = \frac{\text{wt ratio}}{\text{area ratio}}$$

for A:

$$\text{correction factor } F_A = \frac{\text{wt ratio (A/standard)}}{\text{area ratio (A/standard)}}$$

Similar ratios are calculated for B and C.

For the unknown mixture:

5)   Add a sufficient amount of the standard to the unknown to give a peak approximately equivalent in area to the other peaks in the chromatogram. It is not necessary to know the exact amount of the standard added because this value cancels in the ratio calculations that follow.

6)   Obtain a gas chromatogram and determine the area ratio (component/standard) for each component.

for A:

$$\text{area ratio (A/standard)} = \frac{\text{area of A's peak}}{\text{area of standard peak}}$$

7)   Using the correction factor for each component, determine the weight ratio (component/standard).

for A:

$$\text{wt ratio (A/standard)} = F_A \times \text{area ratio (A/standard)}$$

TABLE 12.1   Data for the known sample in the example

Component	Weight (g)	Area (cm²)	Wt ratio (component/ standard)	Area ratio (component/ standard)	Correction factor F (wt ratio/ area ratio)
standard	0.25	6.6	—	—	—
A	0.22	8.30	0.88	1.25	0.70
B	0.26	10.4	1.04	1.57	0.66
C	0.19	5.00	0.76	0.75	1.0

8)   Calculate the weight percent for each component.

$$\text{wt percent A} = \frac{\text{wt ratio (A/standard)}}{\substack{\text{sum of wt ratios of components} \\ \text{(A + B + C) to standard}}} \times 100$$

Because the internal-standard technique involves so many steps, let us go through the procedure with a specific example.

EXAMPLE   **For the known:**

1)   Add 0.25 g of internal standard to a mixture of 0.22 g of compound A, 0.26 g of compound B, and 0.19 g of compound C.

2)   The *areas of the GLC peaks* are listed in Table 12.1.

3)   The *wt ratios and area ratios* (also listed in Table 12.1) for each component/standard are calculated as follows for A:

$$\text{wt ratio (A/standard)} = \frac{0.22 \text{ g}}{0.25 \text{ g}} = 0.88$$

$$\text{area ratio (A/standard)} = \frac{8.3 \text{ cm}^2}{6.6 \text{ cm}^2} = 1.25$$

4)   *The correction factors* (also shown in Table 12.1) for each component in the mixture are calculated as follows for A:

$$F_A = \frac{\text{wt ratio for A}}{\text{area ratio for A}} = \frac{0.88}{1.25} = 0.70$$

For the unknown:

5)   Add about 0.20 g of internal standard to the mixture of unknown composition.

TABLE 12.2   Data for the unknown sample in the example

Component	Area (cm^2)	Area ratio (component/ standard)	Correction factor F^a	Wt ratio (component/ standard
standard	4.0	—	—	—
A	8.0	2.0	0.70	1.4
B	0.75	0.19	0.66	0.12
C	7.2	1.8	1.0	1.8

^{a}From Table 12.1.

6)   The *areas of the GLC peaks* and the *area ratios* (components/standard) are listed in Table 12.2. The area ratios are calculated as follows for A:

$$\text{area ratio (A/standard)} = \frac{8.0 \text{ cm}^2}{4.0 \text{ cm}^2} = 2.0$$

7)   Using the correction factors, determine the wt ratio of each component to the standard. The values obtained are shown in Table 12.2. The calculation for component A follows.

$$\text{wt ratio (A/standard)} = F_A \times \text{area ratio (A/standard)}$$

$$= (0.70)(2.0)$$

$$= 1.4$$

8)   Calculate the *weight percents* from each weight ratio and the sum of the weight ratios.

$$\text{sum of wt ratios of A, B, and C} = 1.4 + 0.12 + 1.8 = 3.32$$

$$\text{wt percent A} = \frac{1.4}{3.32} \times 100 = 42\%$$

$$\text{wt percent B} = \frac{0.12}{3.32} \times 100 = 4\%$$

$$\text{wt percent C} = \frac{1.8}{3.32} \times 100 = 54\% \qquad\qquad \bullet$$

Had the internal-standard technique not been used in the example, and had only the relative GLC *areas* of A, B, and C been used to determine the weight percent values, the erroneous values of 50% A, 5% B, and 45% C would have been obtained.

Suggested Readings

Crippen, R. C. *Identification of Organic Compounds with the Aid of Gas Chromatography*. New York: McGraw-Hill, 1973.

Grant, D. W. *Gas–Liquid Chromatography*. New York: Van Nostrand Reinhold, 1971.

Littlewood, A. B. *Gas Chromatography, Techniques and Applications*. New York: Academic Press, 1962.

Purnell, H. *New Developments in Gas Chromatography*. New York: Wiley-Interscience, 1973.

Condal-Bosch, L. "Some Problems of Quantitative Analysis in Gas Chromatography." *J. Chem. Ed.* **1964,** *41(4),* A235.

Hughes, D. E. P. "Flame-ionization Detector for Gas Chromatography." *J. Chem. Ed.* **1965,** *42(8),* 450.

Karasek, F. W., De Decker, E. H., and Tiernay, J. M. "Qualitative and Quantitative Gas Chromatography for the Undergraduate." *J. Chem. Ed.* **1974,** *51(12),* 816.

Grob, R. L. *Modern Practice of Gas Chromatograph*. 2nd ed. New York: Wiley, 1985.

Perry, J. A. *Introduction to Analytical Gas Chromatograph—History, Principles and Practice*. New York: Marcel Dekker, 1981.

Guiochon, G., and Guillemin, C. L. *Quantitative Gas Chromatography*. New York: Elsevier, 1988.

Problems

12.1 Calculate the retention times of peaks at the following distances from the reference mark if the GLC recorder paper is moving at 3 in./minute.
(a) 12.0 in.   (b) 3.0 cm   (c) 20 mm

12.2 Alcohols have very long retention times on Carbowax columns. Why?

12.3 Suppose an organic compound has a *higher* heat conductivity than the carrier gas. How would its GLC signal appear on a chromatogram run on an instrument with (a) a thermal conductivity detector and (b) a flame ionization detector?

12.4 Why should a thermal conductivity detector exhibit different degrees of response to different compounds?

12.5 Why are the following procedures *invalid*?
(a) To minimize the time required for an analysis, the temperature of a GLC instrument is increased above the boiling points of the components in the mixture.
(b) To increase the retention times and thus maximize the separation between peaks, the gas flow is adjusted to a very low value.

12.6 Between sample injections, some students clean a microsyringe with acetone rather than with the mixture to be analyzed. Why is this a poor technique?

12.7   Helium is the carrier gas of choice for a gas chromatograph containing a thermal conductivity detector, and nitrogen gas is preferred for a gas chromatograph having a flame ionization detector. Suggest a reason for this.

12.8   If the difference in retention times of two components of a mixture is 0.75 minute, by how many centimeters will the signals for the two components be separated on a chromatogram if the chart speed is 2 in./minute?

12.9   How would you prepare a solution of a sample if you want to inject 0.5 μL of sample as a 3-μL injection?

12.10   Suppose that, on a GLC column, methyl hexanoate eluted in 1.87 minutes, methyl heptanoate eluted in 2.35 minutes, and methyl octanoate eluted in 3.0 minutes. What is the structure of an unknown methyl ester, known to belong to the same homologous series, that elutes at 4.7 minutes?

12.11   Using the data given in Table 12.1 (p. 164), calculate the weight percent composition of a mixture with GLC areas of internal standard, 4.3 cm²; A, 6.2 cm²; B, 7.0 cm²; and C, 3.2 cm².

12.12   Determine the weight percentages of the components in mixture I.

Calibration:

Component	Weight (g)	Area (cm²)
standard	0.60	70
A	0.20	25
B	0.45	60
C	0.60	88

Mixture I:

Component	Area (cm²)
standard	30
A	22
B	36
C	18

12.13   In Problem 12.12, if mixture I contains 0.20 g of the internal standard and 0.80 g of the mixture of A, B, and C, what are the actual weights of A, B, and C in the mixture?

Column Chromatography, Flash Chromatography, and HPLC

. . . .

In this chapter, we will discuss three related techniques—*column chromatography*, *flash chromatography*, and *high-pressure liquid chromatography* (HPLC). Column chromatography and flash chromatography are used to separate mixtures for synthetic purposes. HPLC is an analytical instrumental tool utilizing the principles of column chromatography. HPLC is similar to gas chromatography in the manner in which it is used.

. . . .

13.1 Column Chromatography

Column chromatography, also called **elution chromatography,** is used to **separate** mixtures of compounds. In this technique, a vertical glass column is packed with a polar adsorbent along with a solvent. The sample is added to the top of the column; then additional solvent is passed through the column to wash the components of the sample, one by one (ideally), down through the adsorbent to the outlet. Figure 13.1 illustrates the column and the technique.

The sample on the column is subjected to two opposing forces: the solvent dissolving it and the adsorbent adsorbing it. The dissolving and the adsorption constitute an equilibrium process, with some sample molecules being adsorbed and others leaving the adsorbent to be moved along with the solvent, only to be readsorbed farther down the column. A compound (usually a nonpolar one) that is very soluble in the solvent, but not strongly adsorbed, moves through the column relatively rapidly. On the other hand, a compound (usually a more polar compound) that is attracted to the adsorbent moves through the column more slowly.

Because of the differences in the rates at which compounds move through the column of adsorbent, a mixture of compounds is separated into

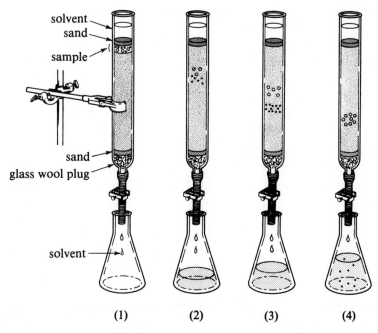

solvent
sand
sample

sand
glass wool plug

solvent

(1)   (2)   (3)   (4)

FIGURE 13.1 Column chromatography. (1) The sample has just started to move into the column of adsorbent. (2) and (3) As more solvent is passed through the column, the sample moves down and begins to separate into its components because of differences in attraction to the adsorbent and solvent. (4) The faster-moving compound is eluted into a flask.

bands, each compound forming its own band that moves through the column at its own rate. Figure 13.1 shows the formation of a pair of bands (which are not always visible). The bands are finally washed out of, or *eluted* from, the bottom of the column, each to be collected in a separate flask. Table 13.1 lists the usual order of elution of different types of compounds.

A. The Adsorbent

The selective action of an adsorbent is very similar to the action of decolorizing charcoal, which selectively adsorbs colored compounds. In fact, activated carbon is sometimes used as an adsorbent in column chromatography. The adsorption process is due to intermolecular attractions, such as dipole–dipole attractions or hydrogen bonding.

$$\underset{\substack{\delta+ \\ \text{dipole–dipole attractions}}}{\overset{\substack{\delta- \quad \delta+ \\ X-R}}{Al_2\,O_3\,\delta-}} \qquad \underset{\substack{Al_2O_3 \\ \text{a hydrogen bond}}}{\overset{\substack{RO \\ | \\ H}}{}}$$

TABLE 13.1 The expected elution order of organic compound classes

	Name of class	General formula
fast	alkanes	RH
	alkenes	R_2C=CR_2
	ethers	R_2O
	halogenated hydrocarbons	RX
increasing polarity	aromatic hydrocarbons	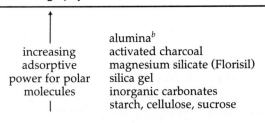 , etc.
	aldehydes and ketones	RCH=O and R_2C=O
	esters	$\overset{O}{\overset{\|}{R C O R}}$
	alcohols	ROH
	amines	RNH_2, R_2NH, R_3N
slow	carboxylic acids	$\overset{O}{\overset{\|}{R C O H}}$

Different adsorbents attract different types of molecules. A highly polar adsorbent strongly adsorbs polar molecules but has little attraction for non-polar molecules such as hydrocarbons. This is why nonpolar compounds are usually eluted first and more polar compounds are eluted later. Because adsorbents differ in their adsorbing power, an adsorbent chosen for a particular chromatographic separation depends in part on the types of compounds being separated. Table 13.2 lists a few of the common adsorbents used to pack chromatography columns.

TABLE 13.2 Typical adsorbents used in column chromatography[a]

↑ increasing adsorptive power for polar molecules ↓	alumina[b] activated charcoal magnesium silicate (Florisil) silica gel inorganic carbonates starch, cellulose, sucrose

[a]Generally, these adsorbents are "activated," usually by heating, to drive off adsorbed water and other volatile substances.
[b]"Basic alumina" contains hydroxide ions. "Acid-washed alumina," with no hydroxide ions, is used for base-sensitive compounds. The adsorptive power of either type of alumina can be adjusted by the addition of a small amount of water.

Silicic acid and alumina are common adsorbents. If one of these two adsorbents is being considered, its ability to separate a particular sample should be checked with various solvents by thin-layer chromatography (TLC; Technique 14). TLC also will indicate the minimum number of compounds in the sample.

The quantity of adsorbent needed depends on the sample, the solvent, and the column. Generally, a minimum of 20–30 g of adsorbent per gram of sample is necessary. This is packed into a column either as a slurry or dry. The column height should be at least ten times its diameter. For the separation of about 0.1–2 g of sample, a column with a diameter of about 3 cm and a height of 30–100 cm is convenient.

B. The Solvent

Generally, solvents are organic compounds. They, too, can be adsorbed by the packing in a chromatography column, and they compete with the sample for positions on the adsorbent. However, nonpolar solvents are not as highly attracted to an adsorbent as are other organic compounds.

The action of a solvent, or a series of solvents, can be used to effect a separation. Assume that you have a mixture of polar and nonpolar compounds and wish to separate the compounds by column chromatography. After adding the sample to the top of the column, you would begin by dripping a nonpolar hydrocarbon solvent like petroleum ether (or a mixture of petroleum ether containing about 1% diethyl ether) through the column. These solvent molecules are not adsorbed to any degree. Polar molecules in the sample are more attracted to the adsorbent than to the nonpolar solvent; therefore, the polar molecules are selectively held back on the column. The nonpolar components of the mixture are not strongly adsorbed and are highly soluble in the nonpolar solvent. These compounds move down the column at a relatively rapid rate to be eluted and collected.

To remove the more polar compounds from the column, you would use a more polar solvent. If you were using a petroleum ether–diethyl ether solvent system, you would gradually increase the percent of diethyl ether so that the polarity of the eluting solvent system would be gradually increased. Table 13.3 lists some common solvents in order of their polarity. Note the similarities between this list and the list in Table 13.1.

The polarity of the solvent should not be changed rapidly, especially when low-boiling solvents are used. The heat generated by interactions of solvents and adsorbents can cause a low-boiling solvent to boil in the column and thus ruin the homogeneity of the column.

C. Collecting the Fractions

In the introductory organic laboratory, chromatographic separations are often carried out with mixtures of colored compounds so that their separa-

TABLE 13.3   Some solvents used in column chromatography

	Name	Structure
increasing polarity	alkanes (petroleum ether, ligroin, hexane)	—[a]
	benzene	C_6H_6
	toluene	$C_6H_5CH_3$
	halogenated hydrocarbons (dichloromethane, chloroform)	CH_2Cl_2, $CHCl_3$
	diethyl ether	$(CH_3CH_2)_2O$
	ethyl acetate	$CH_3\overset{\overset{\displaystyle O}{\|}}{C}OCH_2CH_3$
	acetone	$(CH_3)_2C{=}O$
	alcohols (methanol, ethanol)	CH_3OH, CH_3CH_2OH
	acetic acid	$CH_3\overset{\overset{\displaystyle O}{\|}}{C}OH$

[a]Petroleum ether and ligroin are mixtures of alkanes; see Table 1.1, page 40.

tion on the column can be observed. When one colored component is about to be eluted from the column, a fresh flask should be placed under the column to collect it.

In actual practice, most chromatographic separations involve colorless compounds, whose presence must be detected in other ways. Typically, a series of same-size fractions (25 mL or 100 mL, for example) are collected in tared flasks. It is better to collect many small fractions than a few large ones. These fractions are evaporated and weighed, or they are concentrated and tested by thin-layer chromatography (Technique 14). If TLC is used for monitoring the fractions while the separation is in progress, the operator can check the purity of the compounds as they are eluted and can estimate the number of compounds remaining on the column at any time. Alternatively, HPLC or, sometimes, GLC can be used to monitor the progress of the separation.

• •

SAFETY NOTE   Large volumes of flammable solvents are used in column chromatography, constituting a greater fire and health hazard than usual. Carry out this procedure only in a well-ventilated area, preferably in the fume hood.

• •

D. Steps in Column Chromatography

There are many ways to prepare a column for and carry out a column chromatography; therefore, it is difficult to present a general procedure. We will present only one technique: the preparation of a silicic acid column. Even if your instructor would like you to use a different procedure, we suggest you read through the following steps to learn some of the important features of column packing and use.*

1) Preparing a silicic acid column.   A column must be packed uniformly and contain no holes or air bubbles. The following procedure is one way to prepare a homogeneous wet column.

Sieve the silicic acid; the final particle size should pass an 80-mesh screen but be retained on a 115-mesh screen.† (CAUTION: A dust mask should be worn during the sieving.)

Mount a clean chromatography column (3 cm × 38 cm) as shown in Figure 13.1. Check the column from two directions to make sure it is vertical. With the stopcock (or clamp) closed, pour 50 mL of *n*-heptane (or other nonpolar solvent) into the column. Then, using a glass rod or glass tubing, push a very small plug of glass wool to the bottom of the column. Use the rod to push out any air entrapped in the glass wool. Then add enough sand to the column to cover the glass wool. Using solvent from a dropper, wash down any sand clinging to the sides of the column. Level the sand by tapping the glass column with your finger. Finally, drain about 20 mL of the heptane from the column to ensure that all the air bubbles have been displaced.

Weigh 20 g of sieved silicic acid into a beaker and add 50 mL of heptane. Mix the slurry with a spatula. When completely mixed, the slurry should appear almost translucent and should contain no white lumps of silicic acid or air bubbles. Adjust the stopcock on the column to allow about 5 mL/minute of solvent to pass into the receiving flask. Remix the slurry and pour it in a continuous stream into the chromatography column. Wash the beaker immediately with about 20 mL of heptane and add this wash to the top of the column.

While the heptane is still draining through the column, wash any silicic acid on the sides of the column onto the packing, using heptane from a dropper. Allow the heptane to drain from the column until only about 1 cm remains on top of the packing; then close the stopcock. As the heptane drains from the base of the column, the silicic acid settles and forms the solid support for the chromatographic separation. The packing is very delicate and will be useless if it is allowed to drain dry of solvent.

2) Loading the column.   Allow the heptane to drain from the column until the top of the packing is just free of liquid. Using a dropper, transfer

---

*The procedure outlined here was developed for a separation of 0.10 g of a mixture of the colored compounds *o*- and *p*-nitroaniline (toxic, with potential of dermal absorption)—5% of each in 95% ethanol. (The ethanol was used for ease in column loading.) One additional milliliter of ethanol was used to wash the solution onto the column packing.

†The 80-mesh and 115-mesh values refer to the Tyler Standard Screen Scale. If sieves calibrated by the U.S. Series Designation Standard are used, the silicic acid should pass through a 180-μm screen but be retained on a 125-μm screen.

1 mL of a solution of the mixture to be separated directly onto the top of the packing. Do not let the solution run down the sides of the glass onto the packing. Open the stopcock to allow this solution to flow into the packing. Stop when the top of the packing is just free of liquid.

Measure out no more than 1 mL of solvent and, using a dropper, wash the solution, along with any splatters on the sides of the column, into the packing, again stopping when the top of the packing is just free of the liquid.

Using a clean dropper, add about 3 mL of heptane and drain it into the packing as before, being careful not to disturb the top of the packing. Repeat the process with a second 3-mL portion of heptane. Finally, using a dropper, gently add 10 mL of heptane to the top of the packing; then very carefully pour 90 mL of heptane down the side of the column onto the packing.

3) Developing the column. Adjust the flow rate of the heptane from the column to 1–1.5 mL per minute. (A faster rate of flow will cause the packing to form channels, so that the compounds will "tail," or not move down evenly.) Continue at this rate until about 1 cm of heptane remains on top of the packing.

Carefully add 100 mL of the 3% ethyl acetate–heptane eluting solvent (or other mixture) to the top of the column and continue the eluting process. If the compounds are visible on the column, collect each compound as it elutes from the column in a tared flask. When only one compound remains on the column, a polar solvent can be used to wash it off quickly. If the compounds are not visible on the column, collect 25-mL volumes in tared flasks.

In either case, evaporate the solvent from the flasks. Weigh the flasks to determine the amount of material collected. Thin-layer chromatography (Technique 14) is an excellent analytical tool for analyzing the purity of the compounds you isolate from column chromatography.

Traditional column chromatography, as described here, suffers from a number of practical disadvantages. Primarily, it is time-consuming and tedious; it also requires very large amounts of solvents. Traditional column chromatographic techniques are therefore being replaced by flash chromatography and related instrumental techniques called **high-performance liquid chromatography (HPLC)**. These techniques are discussed in the next sections.

13.2 Flash Chromatography

Classic column chromatography described in the previous section tends to be time-consuming and sometimes results in poor separation because the compound bands have a tendency to tail, especially when the quantities of material being separated exceed 1–2 g. In an effort to overcome these problems, a method known as **flash chromatography** has largely replaced

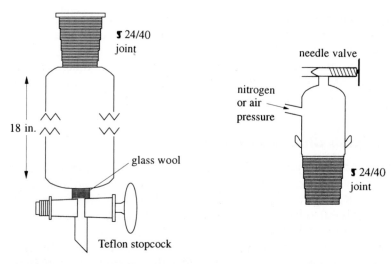

FIGURE 13.2 Apparatus for flash chromatography. (From *Theory and Practice in the Organic Lab* by Landgrebe. Copyright © 1993 by Wadsworth, Inc. Reprinted by permission.)

classical column chromatography. Moderate resolution (equivalent to $R_f \geq 0.15$ on analytical thin-layer chromatography) can be achieved with samples weighing 0.01–10.0 g. The total time required for column packing, sample application, and elution can be as little as 10–15 minutes.

A. Apparatus

The apparatus required consists of a chromatography column 18 in. long, a Teflon® stopcock, and a 24/40 glass joint at the top, as shown in Figure 13.2. The diameter of the glass column depends upon the sample size, as listed in Table 13.4.

TABLE 13.4 Representative Parameters for Flash Chromatographic Separations[a]

Column diameter (cm)	Eluent vol (mL)	Sample loading (mg)[b]		Fraction size (mL)
		$\Delta R_f \geq 0.2$	$\Delta R_f \geq 0.1$	
1	100	100	40	5
2	200	400	160	10
3	400	900	360	20
4	600	1600	600	30
5	1000	2500	1000	50

[a]W. C. Still, M. Kahn, and A. Mitra, *J. Org. Chem.* **1978**, *43*, 2923.
[b]The separation of spots in a thin-layer chromatographic (TLC) analysis using the same adsorbent.
From *Theory and Practice in the Organic Lab* by Landgrebe. Copyright © 1993 by Wadsworth, Inc. Reprinted by permission.

The rate of elution is controlled by air or nitrogen pressure (about 20 psi) supplied to the top of the column. The flow controller is a needle valve also located on the top of the column.

The adsorbent used for flash chromatography has a much smaller particle size than that used for classical column chromatography and is about the same size as that used for thin-layer chromatography.

B. Steps in Flash Chromatography

1) Select a solvent system.   Select a solvent system that gives a good separation and moves the desired component to an $R_f = 0.35$ on a silica gel TLC plate. If this R_f is given by a solvent having less than 2% of the polar component, the solvent selected to elute the flash chromatography column should have about one-half of that percentage.

Solvents found to be especially useful include ethyl acetate–petroleum ether (bp 30°–60°) for general separations and either acetone or methylene chloride with petroleum ether (bp 30°–60°) for separations of polar compounds.

2) Prepare a column.   Select a column of the appropriate diameter (see Table 13.4), put a plug of glass wool at the bottom of the column, add a thin layer of 50–100-mesh dry sand as a platform, and (with the stopcock open) fill it with 5–6 in. of dry 40–60 μm silica gel. Tap the column gently; then add a $\frac{1}{8}$-in. layer of dry sand on top of the silica gel.

Carefully fill the column with solvent, attach and secure the flow controller, and, with the bleed valve open, turn on a small flow of air or nitrogen. Place your finger over the bleed port so that the pressure builds up rapidly in the column and compresses the silica gel. This procedure forces the air pockets out of the bottom of the column. Maintain the pressure until all the air has been expelled; otherwise, the column may fragment and be ruined when the pressure is released.

Release the pressure and readjust the needle valve to maintain a slight pressure so that excess solvent is forced through the column, but do not let the top of the adsorbent go dry. The solvent used in packing the column can be reused to elute the sample.

3) Place the sample on the column.   Introduce the sample, using a Pasteur pipet, as a 20%–25% solution dissolved in the eluting solvent; introduce it to the top of the column and apply a slight pressure to force the sample onto the adsorbent. If the sample is not very soluble in the elution solvent, apply it to the column with a solvent containing a little more of the polar component of the eluting solvent.

4) Develop the column.   Apply adequate pressure to the column to achieve a drop in solvent level of about 2 in./minute. Collect fractions until all of the recommended solvent has been used (Table 13.4).

Because the time required to elute the column is very short (5–10 minutes), it is best to collect the eluent in a rack of at least forty 20 × 150-mm test tubes moved by hand until the separation is complete. It is sometimes advantageous to collect small fractions early and large ones toward the end. Eluted components can be detected by spotting samples from each fraction along the edge of a tall TLC plate and developing the plate sideways. Collected fractions can then be appropriately combined and the solvent evaporated to isolate each component. After the separation has been completed, the silica gel in the column is washed with about 5 in. of ethyl acetate or acetone so that the column can be reused.

·········· 13.3 High-Performance Liquid Chromatography (HPLC)

A relatively new chromatographic analytical procedure is high-performance (or high-pressure) liquid chromatography (HPLC). In HPLC, the solvent is pumped through the column at high pressure (6000 psi; 400 atmospheres). The sample is "injected" into the solvent stream and passes through the column. The components in the sample are separated from one another as they pass through the column. The principles behind their separation are the same as those of column chromatography. As the components in the sample are eluted from the column, they pass through a detector, which converts their passage to electrical potential. The resulting signal is sent to a recorder, which then gives the chromatographic tracing.

The operation of HPLC and the interpretation of the results are similar to those for a gas–liquid chromatography (GLC) instrument (Technique 12). The sample is injected and a tracing is obtained. The number of components in the sample can be estimated by counting the number of peaks in the chromatogram. The amount of each component is proportional to the area under each peak. However, HPLC is a far more powerful tool than GLC and can be used to analyze a mixture that cannot be analyzed with gas chromatography.

A. Instrumentation

A schematic diagram for a high-performance liquid chromatograph is shown in Figure 13.3. The pump, which is constructed from inert materials, forces the mobile phase, or solvent, through the column at a constant flow rate.

Sample injection Because of the high pressure, a simple injection with a hypodermic syringe cannot be used. The high pressure will pop the plunger out the back of the syringe. Therefore, a loop-type injection technique, as diagrammed in Figure 13.4, is used. With the valve in position A (see Figure 13.4), the sample is introduced at atmospheric pressure into the sample loop

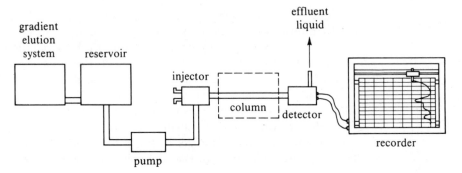

FIGURE 13.3   Diagram of a HPLC instrument. (From *Theory and Practice in the Organic Lab* by Landgrebe. Copyright © 1993 by Wadsworth, Inc. Reprinted by permission.)

of the injector. As the valve is turned toward position B (see Figure 13.4), the flow of the mobile phase in the column is stopped momentarily until the storage loop is properly positioned so that the sample can be swept into the column.

Solvent system   The solvent is stored in a reservoir, which is connected to the pump. In more sophisticated instruments, several different solvents can be stored in reservoirs, and a computer-controlled mixing system can be used for *isocratic* (single-solvent), *stepwise* (sequential-solvents), or *gradient* (gradually changing solvents) elution.

Detection   Different methods have been developed to detect compounds as they elute from the column. These detectors employ conductivity, fluorescence, refractive index, or ultraviolet absorption. Of these methods, the use of ultraviolet absorption is most common because a large number of organic molecules contain function groups that absorb in the ultraviolet region of the spectrum. The method is very sensitive; ultraviolet detectors are capable of

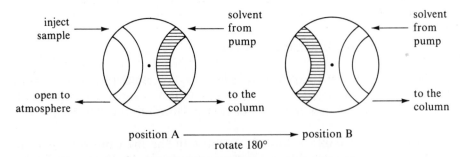

FIGURE 13.4   Loop-type injector. The sample is first injected into the loop that is open to the atmosphere. The injector assembly is then rotated so that the sample is swept by the solvent into the column. (From *Theory and Practice in the Organic Lab* by Landgrebe. Copyright © 1993 by Wadsworth, Inc. Reprinted by permission.)

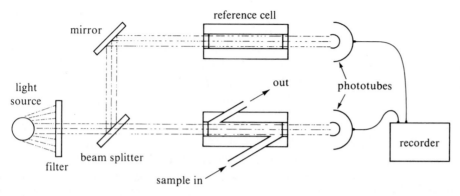

FIGURE 13.5 Diagram of a double-beam ultraviolet HPLC detector. (From *Theory and Practice in the Organic Lab* by Landgrebe. Copyright © 1993 by Wadsworth, Inc. Reprinted by permission.)

detecting 10^{-6}–10^{-10} g of solute. As used in an HPLC instrument, the detector can be single- or double-beam and can employ one or several wavelengths or have a variable-wavelength selector.

Figure 13.5 is a diagram of a double-beam detector. As long as both the sample and the reference cells of the double-beam detector contain only the mobile phase, the electrical output from the two phototubes of the detector is balanced. However, when a solute enters the sample cell, the output from the two photocells becomes unequal; the resulting imbalance is detected and recorded as a chromatogram.

B. Columns

Columns used for HPLC are constructed from stainless steel to withstand high pressures (up to 6000 psi). Cartridge columns come in a variety of sizes but can be as small as 3 cm in length. A short, **guard column** containing the same stationary support as in the main column is used to protect the more expensive main column from being contaminated by impurities that are not easily eluted.

Stationary phases   The most important difference between classical column chromatography and high-performance liquid chromatography is the type of packing used in the column. The classical column is packed with large, irregular-shaped, porous particles (such as silicic acid or alumnia). In contrast, the high-performance column is packed with small beads, called **pellicular particles,** that have a solid core and a thin porous surface layer. A smaller version of these beads is called **microparticles.** In general, pellicular packings have lower efficiencies and sample capacities than microparticles. Pellicular particles, however, can be dry-packed by the chemist in the laboratory. Microparticles must be slurry-packed; and because the technique is quite difficult, prepacked and pretested columns containing microparticles are usually purchased from the instrument manufacturer.

HPLC can be carried out as liquid–liquid or as liquid–solid chromatography. *Liquid–liquid chromatography* is carried out by impregnating an inert solid support with the stationary liquid phase. This type of chromatographic column tends to bleed—that is, the "stationary liquid" tends to be eluted along with the samples being separated.

In an effort to circumvent the problem of bleeding, liquid phases are covalently bonded to the solid support. Chromatography using this type of column is called *liquid–solid chromatography*. Although these stationary phases, or supports, look quite solid in appearance, they behave like a solid particle coated with a liquid.

Stationary liquid–solid supports can be either *normal* or *reverse phase*. In normal-phase chromatography, the stationary liquid phase is more polar than the mobile phase (the solvent). With this type of phase, the less polar compounds are eluted from the column first and the more polar compounds elute more slowly or are tightly adsorbed to the stationary phase. In reverse-phase chromatography, the stationary phase is less polar than the mobile phase (the solvent). Although there are exceptions, in reverse-phase chromatography, the more polar compounds elute first.

Chiral stationary phases   The development and increasing use of chiral stationary phases (CSPs) gives the chemist the possibility of separating a wide variety of enantiomers. Both preparative and analytical separations are possible. The most common use for such CSPs is the determination of optical purity (enantiomeric excess). HPLC columns developed by William H. Pirkle of the University of Illinois are widely used and often referred to as **Pirkle columns.**[*]

Practical considerations in HPLC   Problems associated with the operation of an HPLC instrument often involve some aspect of the pumping–pressure system. Pressures can be high, and only a very small dead volume is permitted in the various fittings and connections in order to preserve any resolution achieved by the column. As a result, the stainless steel tubing that is used has a very small inside diameter (barely visible without magnification) and can become clogged very easily. For this reason, sample solutions and solvents must be free of solid particles larger than a few micrometers in diameter. Special HPLC-grade solvents are used, and water must be freshly purified (usually by ion exchange and reverse osmosis). Even purified water will develop particulate matter (microbial growth) after only a few days unless a small puddle of chloroform is kept in the bottom of the container.

Prior to elution, the solvents should be purged with helium to remove dissolved gases, which can cause bubbles to form in the system, especially

---

[*] W. H. Pirkle, T. C. Pochapsky, G. S. Mahler, D. E. Corey, D. S. Reno, and D. M. Alessi, *J. Org. Chem.* **1986,** *51*, 4991.

after the main column, where the pressure is no longer high. Bubbles passing through the detector create false signals. Even with purged solvents, bubbles can be present throughout the system when the chromatograph has not been used for a while. Although such bubbles can sometimes be purged from the system by a continuous flow of solvent, they can cause the pump to lose its priming and malfunction.

A sudden or gradual buildup of pressure in the system is often a signal that the stainless steel fritted discs (**frits**) are becoming clogged. These frits are placed before and after the guard column and main column to protect the packing, and some instruments have additional frits at other locations. To find the clogged frit requires opening the instrument and testing the pressure at various locations with the pump running.

Suggested Readings

Column Chromatography

Heftmann, E., ed. *Chromatography: A Laboratory Handbook of Chromatographic and Electrophoretic Methods*. New York: Wiley-Interscience, 1971.

Stock, R., and Rice, C. B. F. *Chromatographic Methods*. 3rd ed. London: Chapman and Hall, 1974.

Snyder, L. R., and Kirkland, J. J. *Introduction to Modern Liquid Chromatography*. 2nd ed. New York: Wiley-Interscience, 1979.

Lederer, E., and Lederer, M. *Chromatography. A Review of Principles and Applications*. 2nd ed. New York: Elsevier, 1957.

Cassidy, H. G. "Adsorption Analysis: Tswett's Chromatographic Method." *J. Chem. Ed.* **1939**, *16(2)*, 88.

Karger, B. L. "Resolution in Linear Elution Chromatography." *J. Chem. Ed.* **1966**, *43(1)*, 47.

Flash Chromatography

Still, W. C., Kahn, M., and Mitra, A. "Rapid Chromatographic Technique for Preparative Separation with Moderate Resolution." *J. Org. Chem.* **1978**, *43(14)*, 2923.

Harwood, L. H. "Dry-Column Flash Chromatography." *Aldrichimica Acta* **1985**, *18(1)*, 25.

HPLC

Pryde, A., and Gilbert, M. T. *Applications of High Performance Liquid Chromatography*. New York: Chapman and Hall, 1979.

Parris, N. A. *Instrumental Liquid Chromatography*. 2nd ed. New York: Elsevier, 1984.

Allenmark, S. G. *Chromatographic Enantioseparations—Methods and Applications*. New York: Wiley, 1988.

Clapp, C. C., Swan, J. S., and Poechmann, J. L. "Identification of Amino Acids in Unknown Dipeptides." *J. Chem. Ed.* **1992**, *69(4)*, A122.

Problems

13.1   A chemist wishes to carry out an elution chromatographic separation using diethyl ether and ethanol as the eluting solvents.
(a) With which solvent should the chemist begin the elution?
(b) What would happen if the chemist started with the other solvent?

13.2   A highly polar compound is moving through an elution chromatography column too slowly. What can the chromatographer do to increase its rate of movement?

13.3   Which packing would *not* be suitable for elution chromatography of a mixture of carboxylic acids? Explain your answer.
(a) Basic alumina
(b) A magnesium silicate
(c) Limestone ($CaCO_3$)

13.4   List the following compounds in order of expected elution from a chromatography column packed with acid-washed alumina, using a petroleum ether–diethyl ether solvent system.

TECHNIQUE 14

Thin-Layer
Chromatography

Thin-layer chromatography (abbreviated TLC) is a variation of column chromatography. (If you have not read Technique 13, we suggest that you do so now before continuing with Technique 14.) Instead of a column, a strip of glass, plastic, or aluminum is coated on one side with a thin layer of alumina or silica gel (sometimes mixed with plaster of paris, $CaSO_4$, a binder) as the adsorbent. Other adsorbents can also be used. The quality of the separation for a given mixture depends largely on the adsorbent.

In the TLC analysis, about 10 μL of a solution of the substance to be tested is placed ("spotted") in a single, small spot near one end of the plate using a microcapillary. The plate is "developed" by placing it in a jar with a small amount of solvent. Figure 14.1 shows a TLC plate in a developing jar. The solvent rises up the plate by capillary action, carrying the components of the sample with it. Different compounds in the sample are carried different distances up the plate because of variations in their adsorption on the adsorbent coating. If several components are present in a sample, a column of spots is seen on the developed plate, with the more polar compounds toward the bottom of the plate and the less polar compounds toward the top.

As an analytical tool, TLC has a number of advantages: it is simple, quick, and inexpensive, and it requires only small amounts of sample. TLC is generally used as a qualitative analytical technique, such as checking the purity of a compound or determining the number of components in a mixture or column chromatographic fraction. We can use TLC to follow the course of a reaction by checking the disappearance of starting material and the appearance of product. In addition, TLC is useful for determining the best solvents for a column chromatographic separation. It can also be used for an initial check on the identity of an unknown sample (by spotting the plate with a known compound as well as with the sample). With calibration,

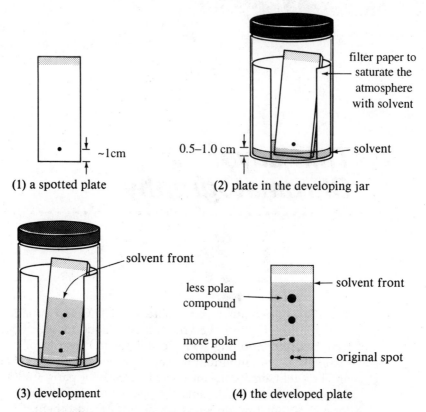

(1) a spotted plate

(2) plate in the developing jar

(3) development

(4) the developed plate

FIGURE 14.1 Thin-layer chromatography. The spotted plate is placed in the developing jar with a piece of filter paper, which acts as a wick to saturate the atmosphere with solvent. Different compounds move up the plate at different rates: the less polar compounds move the fastest and are found closer to the solvent front.

TLC can be used as a quantitative technique. Preparative work can be carried out with special thick-layered TLC plates. TLC is fast, efficient, and simple to use. In all its forms, TLC is a very powerful tool.

· · · · · · · · · · ·
14.1 The R_f Value

The distance that the spot of a particular compound moves up the plate relative to the distance moved by the solvent front is called the **retention factor,** or R_f **value.**

$$R_f = \frac{\text{distance traveled by the compound}}{\text{distance traveled by the solvent}}$$

Figure 14.2 shows how these distances are measured. When the developed

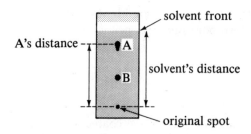

FIGURE 14.2 The R_f value for compound A is the ratio of the distance it has traveled to the distance the solvent has traveled. The spot for A is not circular here but shows "tailing"; therefore, the center of the spot is estimated. (Tailing is usually caused by too much sample in the original spot.)

TLC plate is removed from the developing jar, the solvent front is marked immediately with a pencil before the solvent evaporates. Assuming that the compound spots are colored, the spots are outlined with a pencil in case the color fades. The distance that a compound has traveled is measured from the original spot to the center of the new spot. If the spot is elongated, the "center" is estimated (usually closer to the leading edge). The distance that the solvent has traveled is measured from the original spot to the solvent front.

The R_f value for a compound is a constant only if all variables are also held constant: temperature, solvent, adsorbent, thickness of adsorbent, amount of compound on the plate, and distance the solvent travels. Because it is difficult to duplicate all these factors exactly, an unknown sample is usually compared with a known compound *on the same plate*. Figure 14.3 shows how a mixture containing compound A compares with pure A on the same plate.

If two substances have the same R_f value, they are likely to be (but not necessarily) the same compound. A second TLC comparison using a different solvent for development may result in different R_f values, in which case the substances are not the same. If the second TLC analysis results in identical

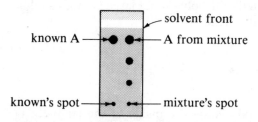

FIGURE 14.3 A known and an unknown sample should be analyzed on the same plate at the same time.

R_f values for the pair, the likelihood that the samples are identical increases. Even in this case, some other type of corroborating evidence is needed.

14.2   Equipment for TLC

1) TLC sheets and plates. Commercial TLC sheets are coated with silica gel (SiO_2) or alumina (Al_2O_3). Choose the type that gives the best separation for your particular mixture.

If commercial TLC sheets are unavailable, plates can be made from microscope slides and a slurry of 1 g aluminum oxide G or silica gel G and 2 mL chloroform (CAUTION: toxic). A 2:1 mixture by volume of dichloromethane and methanol (also toxic) may be substituted for the chloroform.
Dip two slides, back to back, in the slurry. Allow the excess slurry to drain. Separate the slides and allow them to dry in a fume hood. Then wipe excess adsorbent from the backs and edges of the slides. Making satisfactory plates requires practice; therefore, prepare a number of plates and select the most evenly coated ones. Microscope slide plates are shorter than commercial sheets; consequently, the separation of components is not as clean.

2) Pipets. Commercial 10-μL disposable pipets are best for TLC. If commercial pipets are not available, draw out some soft glass tubing or melting-point capillary tubes in a flame. The diameter of the pipet should be about a fourth of the diameter of a melting-point capillary. An excellent spotting pipet can be made by inserting one of these small capillary tubes into a hypodermic needle.

Micropipets cannot be cleaned. Use a fresh pipet for each solution to be spotted, and then discard it after use.

3) Developing jars. Developing jars or chambers with the proper solvent system should be prepared well in advance of their use and kept in the fume hood. Any tall jar with a lid or screw top may be used for a developing jar. The jar should be narrow enough to hold the plate upright inside, without the danger of its falling over (see Figure 14.1). The lid of the jar should be impervious to solvent fumes. The jar should be small enough so that its atmosphere can be quickly equilibrated with solvent but large enough so that the sides of the plate do not touch the wick.

14.3   Steps in a TLC Analysis

1) Preparing the developing jar.   Line the inside of the jar halfway around with a piece of filter paper, which will act as a wick to saturate the atmosphere in the jar with solvent vapor. Before inserting the TLC plate, pour a small amount of the developing solvent into the jar to soak the filter

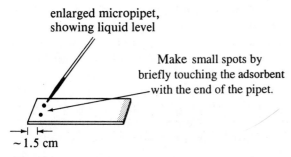

FIGURE 14.4 Spotting a TLC plate with solution from a micropipet.

paper and to cover the bottom of the jar to a depth of about 0.5–1.0 cm. The solvent level should cover the edge of the adsorbent on the plate and yet not reach the spots. Cap the jar and allow it to sit for at least 15 minutes to reach liquid–vapor equilibrium. Check that the solvent level is still about 0.5–1.0 cm, and add more solvent if necessary before inserting the plate. Only one plate can be developed in a jar at a time.

2) Spotting the plate. Dissolve about 1 mg of the solid or liquid sample in a few drops of a volatile solvent such as methanol, CH_3OH, or acetone, $(CH_3)_2C{=}O$. Dip the end of a fresh micropipet into this solution, which will rise into the pipet by capillary action.

At about 1.5 cm from the end of the plate, touch the end of the pipet *gently* and *briefly* to the adsorbent so that the solution runs out of the pipet and onto the adsorbent (Figure 14.4). Do not disturb the coating of adsorbent except where you are spotting. Make the spot as small as possible (1–3 mm in diameter) by allowing only a small amount of liquid to run out before lifting the pipet. As soon as the solvent evaporates, more sample may be added to the same spot. Depending on the concentration of the sample solution, one to three applications are usually sufficient. To determine the optimum number of applications, place three spots on one plate—the first spot containing 1–2 applications, the second spot containing 3 applications, and the third spot containing 4–5 applications.

It is important to spot the compounds high enough on the plate so that they will be *above the solvent level* in the developing jar. If the spots are below the solvent level, they will be dissolved off the plate by the solvent.

If more than one sample is being analyzed on the same plate, space the spots well apart and at the same distance from the bottom of the plate. Samples that are spotted too close together may spread out and run together as they are developed; therefore, a maximum of 3–4 sample spots per 5-cm-wide plate is advised. Use a fresh micropipet for each sample and discard it after use.

3) Developing the plate. Prop the plate upright in the center of the jar (spots at the bottom) in such a way that the adsorbent side of the plate is visible through the side of the jar. Be sure the edges of the plate do not touch the wick. Cap the jar and do not move it during the development.

The solvent will rise up the adsorbent on the plate by capillary action. When the solvent front has risen almost to the top of the plate (about 1–2 cm from the end), open the jar, remove the plate, and quickly mark a line across the plate at the solvent front with a pencil. Check the plate for visible spots. Outline these carefully with the pencil, keeping your lines at the perimeters of the spots. A spot from a colorless organic compound will not be visible on the plate. Therefore, one or more visualization procedures may be followed.

4) Visualizing the spots. The most convenient method of visualizing is with an ultraviolet (black) lamp. Some compounds show up in ultraviolet light as bright spots that then can be outlined. Another technique consists of using a black lamp to visualize spots on a TLC plate containing an inorganic fluorescent compound, such as ZnS, in its coating. When such a plate is placed under a black lamp, the entire plate glows. An organic compound capable of absorbing ultraviolet light, and thus quenching the fluorescence, will show up as a dark spot.

A simple technique for visualizing spots is to place the dry, developed plate in a dry, covered jar with a few crystals of iodine (I_2). (CAUTION: Iodine is a strong irritant!) Most organic compounds show up as colored spots (yellow-brown to purple) when exposed to iodine vapor. The color arises either from the formation of a colored complex between the compound and iodine or from the dissolving of iodine in the compound. In either case, vapors of iodine are selectively adsorbed onto the TLC plate wherever there is a concentration of organic compound. If spots do not appear, warm the bottom of the jar gently (with your hand or briefly on a steam bath) to vaporize the iodine crystals. As soon as the spots are well defined, remove the plate from the jar and circle the spots with a pencil. This step is necessary because, as the iodine rapidly sublimes from the plate, the spots will become colorless again.

If a plate is left in the iodine vapor too long, the entire plate will become dark, and the spots will no longer be distinguishable. In this case, remove the plate and observe it carefully as the iodine sublimes. Generally, the iodine will evaporate more quickly from the bulk of the plate than from the organic compounds. The spots should thus become apparent, and they can be circled with a pencil.

Other valuable techniques for spotting involve *charring* and the use of *color-forming reactions.* If the backing of the plate is glass or aluminum (but not plastic), the organic compounds can be visualized by heating the plate to high temperature in an oven. Often, the plate is sprayed with sulfuric acid to hasten the charring.

An alternative, and often very specific, procedure involves spraying the plate with a reagent that will react with a compound on the plate to form a colored product. For example, an acid–base indicator, like phenolphthalein, can be used to identify acidic or basic compounds. Other specialized sprays contain ferric chloride (to identify phenols), 2,4-dinitrophenylhydrazine, DNP (to identify aldehydes and ketones), and, with paper chromatography, ninhydrin (to identify amino acids).

· · · · · · · · · · ·
14.4   A Related Technique: Paper Chromatography

In many respects, **paper chromatography** is similar to thin-layer chromatography. Instead of an adsorbent-coated plate, a strip of paper is used. Instead of a solid adsorbent, a thin film of water on the paper constitutes the adsorbent. Therefore, paper chromatography is a *liquid–liquid* partition technique, rather than a liquid–solid technique such as column chromatography and TLC.

For exacting analyses, commercial chromatographic paper strips equilibrated in a humid atmosphere should be used. In many analyses, however, filter paper can be used for paper chromatography because it is almost pure cellulose with few impurities. Under most atmospheric conditions, filter paper adsorbs moisture from the air. This adsorbed water makes up about 20% by weight of the filter paper and is usually sufficient for successful paper chromatography. Often, a damp solvent is used to develop the paper chromatogram to ensure the presence of sufficient water. A solvent that is fairly immiscible with water does not disturb the film of water adsorbed onto the cellulose.

Because very polar water molecules form the adsorbent layer, paper chromatography is most successful with very polar organic compounds. (Nonpolar compounds, which are not attracted to water, are usually carried with the solvent front.) Paper chromatography is commonly used for the identification of amino acids, which exist as highly polar *dipolar ions,* species containing full positive and negative ionic charges.

$$H_3\overset{+}{N}-\underset{\underset{R}{|}}{\overset{\overset{H}{|}}{C}}-\overset{\overset{O}{||}}{C}O^-$$

the dipolar ion
of an amino acid

Suggested Readings

Stahl, E., ed. *Thin-Layer Chromatography. A Laboratory Handbook.* 2nd ed. New York: Springer, 1969.

Hais, I. M., and Macek, K., eds. *Paper Chromatography.* New York: Academic Press, 1963.

Batt, C. W., and Kowkabany, G. N. "Distance Factors in Paper Chromatography." *J. Chem. Ed.* **1955,** *32(7)*, 353. (The principle also applies to TLC.)

Bobbit, J. M. *Thin-Layer Chromatography.* New York: Reinhold Publishing Corp., 1963.

Randerath, K. *Thin-Layer Chromatography*. 2nd ed. New York: Academic Press, 1966.

Kirchner, J. G. *Technique of Organic Chemistry. XII. Thin-Layer Chromatography.* A. Weissberger and E. S. Perry, eds. New York: Interscience Publishers, 1967.

Geiss, F. *Fundamentals of Thin-Layer Chromatography*. New York: Springer-Verlag, 1987.

Kirchner, J. G. *Thin-Layer Chromatography*. 2nd ed. New York: Wiley-Interscience, 1978.

Touchstone, J. C., and Dobbins, M. F. *Practice of Thin-Layer Chromatography*. 2nd ed. New York: John Wiley and Sons, 1983.

Problems

14.1   Calculate the R_f values for the following compounds.
(a) Spot, 5.0 cm; solvent front, 20.0 cm
(b) Spot, 3.0 cm; solvent front, 12.0 cm
(c) Spot, 9.8 cm; solvent front, 12.0 cm

14.2   If two compounds have R_f values of 0.50 and 0.61, how far will they be separated from each other on a plate when the solvent front is developed to (a) 5 cm, and (b) 15 cm?

14.3   As a separation and detection method, would TLC or paper chromatography yield better results in the analysis of each of the following pairs of compounds?

(a) CH_3CH_2SOH and $CH_3CH_2SNH_2$
(with two O double bonds on each S)

(b) $H_2N-\langle O \rangle-CO_2H$ and   (with H_2N on ring) $\langle O \rangle-CO_2H$

(c) $CH_3(CH_2)_6CCH_3$ and $CH_3(CH_2)_6CHCH_3$
(with O double bond on C)   (with OH on C)

14.4   A wick of filter paper is placed in a TLC developing jar, and the atmosphere in the jar is saturated with solvent before a plate is developed. What would happen if a plate were developed in a jar with an atmosphere not saturated with solvent vapor?

14.5   Would you expect Br_2 or Cl_2 to be as suitable as I_2 for a visualizing agent in TLC? Explain.

14.6   Which of the following reactions could be used as the basis for a visualization technique?
(a) $R_2C{=}CR_2 + I_2 \rightleftharpoons R_2CI{-}CIR_2$
(b) $RCO_2H + NaOH \longrightarrow RCO_2Na + H_2O$
(c) $RH + H_2SO_4 \longrightarrow C + CO_2 + H_2SO_4 \cdot H_2O$

14.7   Like ZnS, some organic compounds are fluorescent. Could one of these compounds be used in place of ZnS as a visualizing agent in TLC? Explain.

TECHNIQUE 15

Carrying Out
Typical Reactions
···

In the laboratory, an organic chemist often conducts an experiment or a synthesis by the following series of steps:

1) Carry out the reaction.
2) Isolate the product, perhaps by extraction, crystallization, distillation, or a combination of these techniques.
3) Purify the product, perhaps by crystallization, distillation, sublimation, or a chromatographic technique.
4) Characterize the product and estimate its purity by one or more of the following techniques: melting point, boiling point and refractive index, gas–liquid chromatography, thin-layer chromatography, spectroscopy, or elemental analysis.

Steps (2) and (3), isolating and purifying the product, are often referred to as the **workup** of the reaction mixture. How a reaction mixture is worked up depends on the physical and chemical properties of the product and by-products. Many of the techniques used to isolate, purify, and characterize an organic compound have already been presented. In this technique, we will discuss only Step (1) in the preceding list: how typical organic reactions are carried out. Our emphasis will be on macroscale reaction (microscale reactions were discussed in Technique 10).

An organic reaction is carried out in a *reaction vessel*, generally a flask (Erlenmeyer, round-bottom, or three-neck). Choosing and equipping the vessel depend on the reaction itself, which may require stirring, heating or cooling, and addition of reagents during the reaction's course. Therefore, let us consider these operations in light of the equipment required.

. . . .
15.1   Stirring

Homogeneous reaction mixtures need not be stirred, even when heated, because the reactants are in intimate contact. However, to prevent bumping, boiling chips must be added to a homogeneous unstirred mixture before it is heated to boiling. Heterogeneous reaction mixtures, or mixtures that may become heterogeneous, require stirring to mix the reactants. Stirring is particularly important when a heterogeneous mixture is heated, because stirring prevents bumping. (Boiling chips do not prevent bumping of heterogeneous mixtures containing solid materials).

The simplest stirring technique is *swirling*. Swirling should be used only to mix solutions at, or near, room temperature. A flask containing a boiling liquid should not be swirled, because hot vapors from the flask can burn the hand. When the reaction vessel is an Erlenmeyer flask or beaker, another simple stirring technique is hand stirring with a *glass rod* (not a thermometer).

A **magnetic stirrer,** used with a beaker or flask, accomplishes simple mixing mechanically. A stir bar, generally a ferrous metal encased in an inert Teflon® coating, is placed in the reaction vessel. The vessel is then placed on the motor casing. When the motor is turned on, it rotates a magnet, which causes the stir bar to turn (see Figure 15.1). The principal disadvantage of magnetic stirring is that sufficient torque cannot be generated for stirring a thick mixture.

Some magnetic stirrers are equipped with hot plates so that mixtures can be warmed and stirred simultaneously. When heating a flask in a water bath while using one of these devices, you must use a nonferrous pan

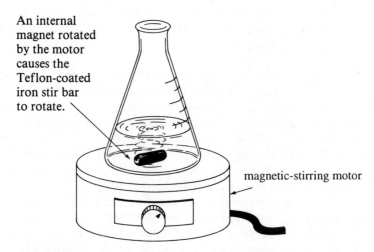

An internal magnet rotated by the motor causes the Teflon-coated iron stir bar to rotate.

magnetic-stirring motor

FIGURE 15.1   A magnetic stirrer can be used to stir mixtures that are not viscous.

(aluminum or glass). Similarly, a cooling bath used with a magnetic stirrer must be nonferrous.

The stirrer of choice is a **mechanical stirrer** powered by a sparkproof motor. This type of stirrer, which is commonly found in research laboratories but rarely in student laboratories, is sufficiently powerful to stir even viscous mixtures. Figure 15.5 (page 198) depicts a mechanical stirrer. The stir blade at the bottom of the heavy glass rod is made from glass or Teflon® and can be changed to fit the size of the flask.

15.2 Heating Reaction Mixtures

Most organic reactions require that the reactants be heated. As in distillation, the heat source chosen depends on the temperature that must be attained and the flammability of evolved vapors. Thus, a steam bath, a hot plate, a heating mantle, a heat lamp, or (rarely) a burner might be chosen. For a reaction that must be kept very dry, a heating mantle is preferable to a steam

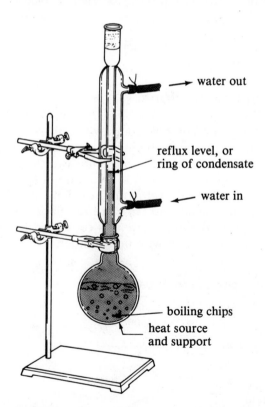

water out

reflux level, or
ring of condensate

water in

boiling chips
heat source
and support

FIGURE 15.2 A round-bottom flask fitted with a reflux condenser.

bath. Regardless of the heat source, the heat should be applied evenly to avoid localized overheating.

Heating under reflux Reflux means "a flowing back." In the chemistry laboratory, the term *reflux* refers to the return of condensed vapors to their original vessel.

Figure 15.2 shows a reaction assembly for heating a mixture under reflux. The **reflux condenser** is an ordinary condenser arranged in an upright position so that vapors from the boiling liquid are condensed and returned to the flask. A reflux condenser can be used during spontaneous exothermic reactions or when a liquid is being boiled by a heat source. The purpose of reflux is twofold: a reaction can be maintained at the temperature of the boiling solvent, and solvent is not lost to the atmosphere.

When heating a compound under reflux, keep in mind that heating too strongly will drive solvent vapors out the top of the condenser. A boiling reaction mixture is no warmer, nor will the reaction proceed any faster, when unnecessary heat is applied. Therefore, control the rate of heating so that the reflux level is in the *lower* portion of the condenser. If a low-boiling solvent is being heated, a second reflux condenser, placed on top of the first one, helps control the escape of vapors.

· · · · · · · · · · ·
15.3 Controlling Exothermic Reactions

Many reactions in your laboratory course are *exothermic reactions,* reactions that produce heat. If a reaction is highly exothermic, its temperature must be controlled. Otherwise, the reaction can "run away"—that is, proceed too rapidly and boil too vigorously, even to the point of spewing out the top of the condenser! A few typical exothermic reactions are oxidation reactions, reactions using sodium metal, and Grignard reactions.

The rate of an exothermic reaction is controlled in three ways, which are usually used in conjunction with one another.

TABLE 15.1 Some common cooling baths

Cooling bath	Minimum temperature (°C)
ice–water	0
ice–salt–water (100 g ice–33 g NaCl)	−21
ice–calcium chloride (70 g ice–100 g $CaCl_2 \cdot 6H_2O$)	−54
dry ice–ethanol	−72
dry ice–acetone	−77

1) *Regulation of the temperature,* usually by using a pan of ice water to chill the reaction flask. (You should *always* have an ice bath at hand when running an exothermic reaction.) Table 15.1 lists some other cooling baths and their minimum temperatures.

2) *Regulation of the rate of addition* of one of the reactants. (Techniques for controlling the rate of addition are discussed in Section 15.4.)

3) *Continuous stirring or swirling* during the addition so that "pockets" containing an excess of a reactant do not develop. (Thorough mixing is especially important if the reaction mixture is thick because of precipitated solid.)

15.4 Adding Reagents to a Reaction Mixture

In many reactions, the reactants are simply mixed and the mixture is heated under reflux until reaction is complete. In many other reactions, one reactant must be added slowly to the reaction mixture without allowing noxious fumes and vapors to escape.

A. Addition of Liquids

Small amounts of a liquid can be added to a reaction mixture by dropping them down a reflux condenser. The preferred technique, however, is to add the liquid from a dropping funnel. Figure 15.3 shows a simple reaction assembly equipped with a dropping funnel. (This figure also shows a *drying tube,* through which air can pass; see Section 15.5.) Note that the reaction flask is topped with a Claisen head and that the dropping funnel is positioned directly above the reaction flask. This positioning allows liquid to be added directly to the reaction mixture without its running down the sides of the glass. The other arm of the Claisen head is connected to a reflux condenser so that vapors can be condensed and returned to the reaction flask. A dropping funnel cannot be placed directly atop the reflux condenser as shown in Figure 15.2, because this would constitute a closed system.

B. Addition of Solids

Adding solids to a reaction mixture is generally more difficult than adding liquids. If the solid is granular (not powdery) and not likely to stick to the wet inner sides of a reflux condenser, it may be feasible to drop the solid directly through the reflux condenser. However, usually a solid must be added directly to the reaction mixture without coming into contact with a glass surface. One way to add solids is to cool the reaction mixture, remove the reflux condenser, and add the solid to the flask itself. Alternatively, set up the reaction assembly shown in Figure 15.3, but with a ground-glass

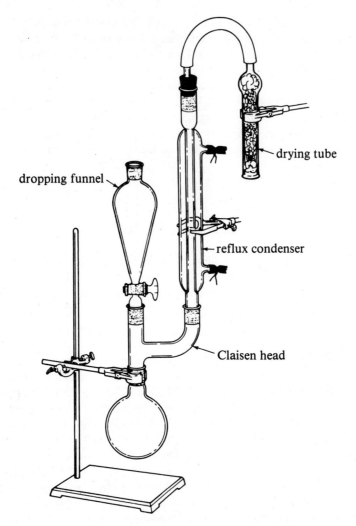

FIGURE 15.3 A reaction setup that allows a liquid to be added to a mixture under reflux.

stopper in place of the dropping funnel. Then, the solid can be added through this joint. When adding a solid through a ground-glass joint, always use a *powder funnel*. The wide stem will allow the solid to pass without clogging and will protect the ground-glass joint from contamination.

15.5 Excluding Moisture from a Reaction Mixture

Atmospheric moisture tends to condense inside a cold condenser. Some organic reactions require that moisture be excluded from their atmosphere. In a research laboratory, a chemist might exclude both oxygen and moisture

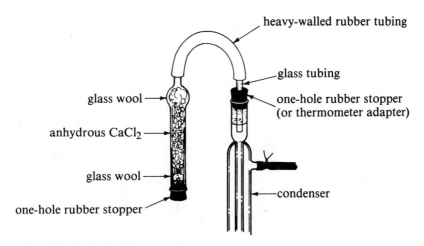

FIGURE 15.4 How to attach a straight-neck drying tube to the reaction assembly (clamp not shown).

by running a reaction under an atmosphere of a dried inert gas such as nitrogen or argon. This is seldom feasible in a student laboratory. Instead, a **drying tube** (Figure 15.4), containing coarse pieces of a desiccant, is used. Anhydrous calcium chloride is the most common desiccant for this purpose.

The purpose of the drying tube is to prevent atmospheric moisture from entering the reaction vessel via the condenser and yet allow the reaction vessel to be open to the atmosphere so that gas pressure does not build up. There are two types of drying tubes: curved (better) and straight (less expensive). A straight drying tube must not be connected directly to the top of the condenser because the desiccant can liquefy and drain into the condenser. Connect the straight tube to the condenser by a short length of heavy-walled rubber tubing, as shown in Figures 15.3 and 15.4.

In either type of drying tube, the desiccant is held in place with loose plugs of glass wool. A one-hole rubber stopper may be used as a secondary plug at the wide end of the drying tube.

The length of time the desiccant in a drying tube remains sufficiently dry to be effective depends on the relative humidity and on the reaction vapors to which it is subjected. You cannot judge the dryness or wetness of calcium chloride by its appearance; however, desiccants containing moisture indicators are commercially available. (Drierite® is anhydrous calcium sulfate impregnated with an indicator that is blue when dry and pink when wet.) A small amount of such a desiccant added to the anhydrous calcium chloride in a drying tube will indicate when the calcium chloride should be replaced.

· · · · · · · · · · ·
15.6 Setting Up a Three-Neck Reaction Flask

Your laboratory may not be supplied with two- or three-necked flasks for your experiments; however, they are the standard reaction vessels in re-

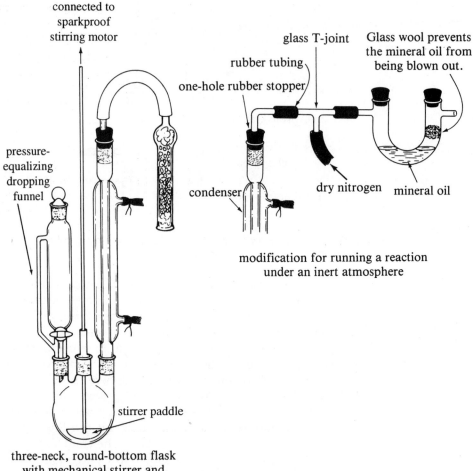

connected to
sparkproof
stirring motor

glass T-joint

Glass wool prevents
the mineral oil from
being blown out.

rubber tubing

one-hole rubber stopper

pressure-
equalizing
dropping
funnel

condenser

dry nitrogen   mineral oil

modification for running a reaction
under an inert atmosphere

stirrer paddle

three-neck, round-bottom flask
with mechanical stirrer and
pressure-equalizing dropping funnel

FIGURE 15.5 A three-neck flask reaction assembly (clamps not shown). Care should be taken that solvent fumes do not flow over the stirring motor.

search and advanced organic laboratories. These flasks are convenient for reactions that require more than one attachment to the reaction flask.

Figure 15.5 shows a three-neck flask arrangement that would be suitable for a Grignard reaction and many other types of organic reactions. Note the *mechanical stirrer*, which is set in the center joint. The stirrer column must be carefully aligned so that it does not bind. The *reflux condenser* is set into a side joint of the flask. The dropping funnel shown is a *pressure-equalizing dropping funnel*. Its sidearm is connected to the atmosphere within the flask (dry air and solvent fumes) so that its contents can be isolated from atmospheric moisture without pressurizing the funnel. An ordinary dropping funnel equipped with a drying tube on top would accomplish the same purpose.

Figure 15.5 also shows how the three-neck flask apparatus can be modified for running a reaction under an inert atmosphere (dry nitrogen or argon). After the air is swept from the reaction apparatus, the inert gas is kept at a slight positive pressure. The apparatus is not continuously swept with gas because solvent vapors would be carried into the room. The mineral oil bubbler excludes air without sealing the system.

15.7 Hydrogen Halide Gas Traps

Reactions in which gaseous HCl and HBr are given off should be carried out in the fume hood. Unfortunately, many laboratories do not have sufficient hood space for an entire class. In such a situation, the HCl and HBr must be trapped.

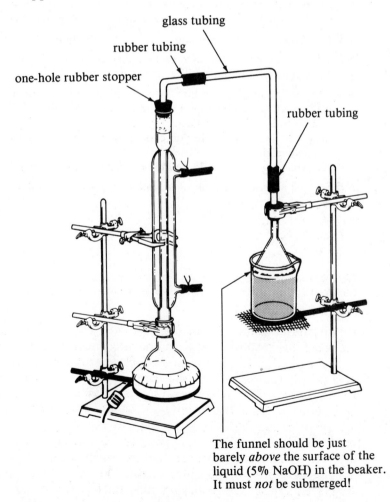

The funnel should be just barely *above* the surface of the liquid (5% NaOH) in the beaker. It must *not* be submerged!

FIGURE 15.6 A gas trap for HBr or HCl.

Figure 15.6 shows a simple trap for a hydrogen halide. The beaker contains 5% aqueous NaOH (caustic, even at this dilution). The funnel is clamped so that it is *just above* the alkaline solution. (If the funnel is submerged, a pressure drop in the reaction vessel could cause the alkaline solution to be drawn into the reaction flask, with possibly disastrous consequences.)

Problems

15.1 Suggest reasons for the following procedures when you are setting up a reaction assembly.
(a) The ground-glass joints are lightly greased.
(b) A gentle reflux is used rather than a vigorous reflux.
(c) Only a small amount of indicator desiccant is placed in a drying tube, instead of the tube being filled completely with this material.
(d) Powdered drying agents are not used in drying tubes.

15.2 List three ways of preventing a "runaway" reaction.

15.3 When HCl is emitted from a reaction mixture, why do we *not* simply bubble the effluent gas through a solution of NaOH to trap it?

15.4 The experimental procedure in a chemical journal describes the exothermic reaction of 10.0 g of compound A in 25 mL of solvent with 75 mL of solution B in a 250-mL Erlenmeyer flask.
(a) What size flask would you choose for the reaction of 500 mg of compound A? Explain.
(b) What size flask would you choose for the reaction of 15.0 g of A? Explain.

15.5 Suggest a reason why nitrogen, rather than argon, is more commonly used as an inert atmosphere.

Sample Preparation for Spectroscopy

The theory of spectroscopy and the techniques for interpreting spectra can be found in any adequate organic textbook written for the yearlong course. We will not attempt to cover those topics here. Our goal will be to discuss more practical features of preparing samples for infrared (IR), nuclear magnetic resonance (NMR), and ultraviolet (UV)/visible spectroscopy.

16.1 Preparing a Sample for an Infrared Spectrum

A **sample cell** is a sample container that fits into the spectrophotometer in the path of the infrared radiation. Different types of cells are available for solids, liquids, solutions, and even gases. The "windows" of most common infrared cells, which allow the infrared radiation to pass through the sample, are polished sodium chloride plates. Sodium chloride is used because it is transparent to infrared radiation in the region of interest.

Sodium chloride plates and cells are relatively expensive because each plate is cut from a single giant crystal and then polished. The plates and cells are also fragile and sensitive to moisture. Atmospheric moisture, as well as moisture in a sample, causes the polished plate surfaces to become pitted and fogged. Fogged plates scatter and reflect infrared radiation instead of transmitting it efficiently; poor-quality spectra are the result. Fogged plates may also retain traces of compounds from previous runs, giving rise to false peaks. If it is necessary to use fogged plates, be sure to clean them carefully, then run a blank spectrum of the plates without sample before running the spectrum of your sample. By doing this, you can determine if your spectrum will contain extraneous peaks.

To prevent fogging, use only scrupulously dried samples for spectral work. Handle the plates *only by their edges*, never by their flat surfaces, because moisture from your fingers will leave fingerprints. NaCl plates and solution cells must be cleaned with a dry solvent such as CH_2Cl_2 after use and stored in a desiccator.

Fogged thin-film plates can be polished by rubbing them with a circular motion on an ethanol–water-saturated paper towel laid on a hard surface. Check with your instructor before you attempt to polish a plate!

A. Liquid Samples

Liquid samples are usually analyzed *neat* (meaning "pure" or "without solvent") as **thin films.** A drop or two of the liquid is sandwiched between two NaCl plates; then the plates are mounted in a holder and placed in the spectrophotometer. Figure 16.1 illustrates this technique.

Few problems are encountered in thin-film sampling, and these are easily **solved.** *Too much liquid* between the plates gives rise to a spectrum in which many of the peaks are too strong—in the 0%–10% transmission range, or even "off the paper." Also, leakage around the edges of the plates can contaminate the plate holder. To remove the excess liquid, wipe part of the sample off the plates (gently) with a dry tissue. Then, rerun the spectrum.

Too little sample between the plates results in a spectrum with weak peaks. More sample may be added to the plates and the spectrum rerun. If a spectrum shows acceptably strong peaks at the start, followed by progressively weaker peaks, the sample may be evaporating. One solution to this problem is to stop the scanning during the run, put more sample between the plates, then continue the scan. However, a preferred technique for sampling a volatile liquid is as a solution.

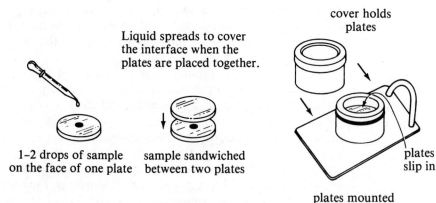

Liquid spreads to cover the interface when the plates are placed together.

cover holds plates

1–2 drops of sample on the face of one plate

sample sandwiched between two plates

plates slip in

plates mounted in holder

FIGURE 16.1 Preparing a liquid sample for infrared spectroscopy by the thin-film technique.

B. Solid Samples

A solid sample must be made transparent to infrared radiation before its spectrum is run. Three ways to do this are to include the sample in (1) a mull, (2) solution, or (3) a KBr pellet.

1) A **mull** is prepared by grinding the solid sample with an inert carrier (petroleum jelly or mineral oil) to the consistency of toothpaste or thick gravy. The mull is then sandwiched between NaCl plates and run as a liquid thin film.

The older (and less satisfactory) technique for preparing a mull is to grind the sample with mineral oil in an agate mortar. It is important to grind the sample very thoroughly before and after adding the oil. Even moderately large solid particles in the thin film cause reflection and scattering of the infrared radiation.

A proper balance of mineral oil to sample is surprisingly difficult to achieve and is best determined experimentally. Begin by grinding together about 10–20 mg of solid with 2–3 drops of mineral oil. Use the spectrum to show if more sample or more mineral oil is needed. If the ratio of sample to mineral oil is correct, the strongest sample (not mineral oil) peaks will dip to about 40%–50% transmission.

A faster way to mull a sample with mineral oil is by grinding them between two pieces of ordinary glass whose inner surfaces have been ground with carborundum. [See C. E. Brady, *J. Chem. Ed.* **1969**, *46(5)*, 301.]

Alternatively, use the ground-glass joint of a flask and a stopper for the grinding. Place a little petroleum jelly (such as Vaseline) on the inside of the joint and sprinkle 0.2–0.3 g of the solid on the surface of the jelly. Grind the sample by twisting the ground-glass stopper and pressing the stopper and joint together. Occasionally, remove the stopper, scrape up the material that has oozed out with a metal spatula, return it to the center of the joint, and then continue grinding. The entire procedure should take no more than 3 minutes. (Mineral oil is not viscous enough to be used in this way.)

Mineral oil and petroleum jelly are mixtures of hydrocarbons and consequently show hydrocarbon absorption bands in the infrared spectrum. Before running a mull spectrum, first run the spectrum of a thin film of mineral oil or jelly alone for reference. By doing this, you will be aware which absorption bands in your sample spectrum arise from the carrier. Some workers place a thin film of mineral oil or jelly in the reference beam of the spectrophotometer to block out its absorption from the sample's spectrum. (The instrument automatically subtracts absorption of the reference beam from absorption of the sample beam.) This procedure produces a cleaner-looking spectrum but decreases the sensitivity in the CH and C—C regions of the spectrum.

After running a mull spectrum, clean the NaCl plates gently with dry solvent and a tissue before returning them to the desiccator.

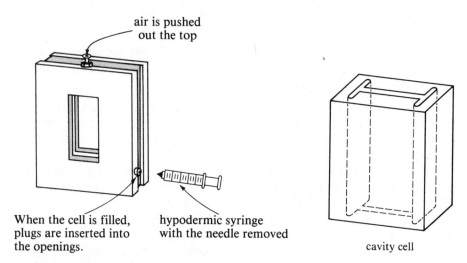

2) A **solution spectrum** is obtained by preparing a 5%–10% (by weight) solution of the sample in a suitable solvent and placing the solution in an infrared **solution cell** (Figure 16.2). A matched cell filled with solvent alone is placed in the reference beam to compensate for solvent absorption bands. Because the solvent bands are not always entirely subtracted, you should be aware of the solvent's absorption frequencies. As with mineral oil or petroleum jelly, placing solvent in the reference beam decreases the sensitivity of the spectrum in the regions where the solvent absorbs radiation. If you do not know what concentration of solution to use, run a trial spectrum on a 10% by weight solution; then add more solvent or sample as needed to obtain a good spectrum.

By far the simplest cell is the cavity cell (Figure 16.2), which consists of a single block of sodium chloride with a 0.1-mm slit machined down the center of the block. These cells require only 0.05 mL (50 μL) of solution, which means that suitable spectra can be obtained with as little as 2–3 mg of sample. The cavity cell can be filled by delivering about 50 μL of a solution with a Pasteur pipet to the slit at the top of the cell.

The solvent chosen must dissolve the sample and must not contain any functional groups that would interfere with the spectrum of the sample. Typical infrared solvents are CCl_4, $CHCl_3$, and CS_2. Of these, CCl_4 is the most useful because its only absorption is at 700–800 cm^{-1} (12–14 μm), at the extreme right of the spectrum. The other solvents show some absorption in the more useful regions of the infrared spectrum, but one of these solvents may be necessary if the sample is insoluble in CCl_4. Commercial CCl_4 and $CHCl_3$ contain ethanol as a stabilizer and must be freshly distilled. In

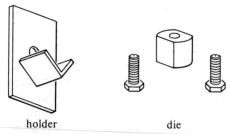

holder   die

FIGURE 16.3   KBr pellet holder and die.
(From *Theory and Practice in the Organic
Lab* by Landgrebe. Copyright © 1993 by
Wadsworth, Inc. Reprinted by permis-
sion.)

addition, CCl_4, $CHCl_3$, and CS_2 are all toxic. Carbon disulfide presents the
added hazard of an extremely low ignition temperature. A hot plate, a steam
bath, or even a radiator pipe can ignite CS_2 vapors. Safe handling procedures, such as transferring these solvents only in a fume hood, should be
observed.

After carrying out a solution spectrum, flush the cell thoroughly with
the solvent. Dry the cell by placing a syringe barrel filled with desiccant in
one port of the cell and drawing air through the cell with a second syringe
attached to the other port. Then return the cell to the desiccator. To clean a
cavity cell, hold it tightly by the edge and flick it into a hood to remove the
solution. Add fresh, dry solvent to fill the cell and flick it out. Repeat this
operation several times.

3) A **KBr pellet** or **wafer,** is obtained by putting great pressure on a
mixture of finely ground sample (dry!) and specially dried potassium bro-
mide until a clear or translucent pellet is formed. Either a hydraulic press or a
special die, tightened by wrenches, must be used to prepare the pellet.

A die used in many student laboratories is simply a pair of stainless
steel bolts with flat polished ends and a nut (see Figure 16.3). To prepare a
KBr pellet using this die, finely grind 1–2 mg of sample with 100 mg of
anhydrous KBr (stored in the oven) in an agate or glass mortar. Screw one
bolt of the die into the nut. Pour about one-half of the sample mixture into
the other end and tap the nut to level the sample. Screw in the second bolt to
sandwich the sample in the middle. Place the die into a holder designed to
prevent one bolt from turning. Tighten the other bolt with a torque wrench
to about 20 ft-lb. After 30–60 seconds, remove the bolts.

Leave the pellet in the nut, place the nut in the special infrared sample
holder, and run the spectrum. Alternatively, the pellet can be removed from
the die and placed on the holder (see Figure 16.3) in the beam. When the
spectrum is complete, remove the pellet from the nut. Wash and dry the die
pieces and store them in the oven.

• • • • • • • • • • •

16.2 NMR Spectroscopy

A. Preparing a Solution for a Proton NMR Spectrum

In some laboratories, students are expected to operate the NMR spectrometer; in other laboratories, students submit prepared samples to a spectrometer operator. Your instructor will specify the procedure used in your laboratory and will instruct you in the use of the instrument if you are to obtain your own spectra.

An NMR spectrum is usually carried out with a solution of the sample in a suitable solvent. The solution, along with a small amount of TMS, is placed in the sample container, called an **NMR tube**, which is capped and placed in the instrument. The sample tube is rotated rapidly, spun by an airstream in the instrument. This spinning results in an averaged magnetic field throughout the entire sample.

A typical NMR tube holds 0.4–0.5 mL of solution. The concentration of sample is usually 10%–30%.

The best solvents for NMR spectra are those that contain no protons, such as CCl_4, CS_2, or $CDCl_3$. For water-soluble compounds, D_2O may be used; however, TMS is water-insoluble and a different internal standard must be used. Almost all deuterated solvents contain traces of nondeuterated solvent. Therefore, a background spectrum of the solvent should be run before the sample is dissolved in it. Deuterated solvents are very expensive and must not be wasted. Solubility of a sample should be tested in the nondeuterated analogue, such as $CHCl_3$ or H_2O, before a deuterated solvent is used.

All NMR solvents must be extremely pure. Reagent-grade CCl_4, for example, is not suitable for NMR spectra because it contains ethanol added as a stabilizer. If there is any doubt about the suitability of a solvent for NMR work, run a background spectrum of the solvent plus a drop or two of TMS.

All solid material must be removed from the solution before it is placed in the NMR tube. Suspended or solid particles (particularly magnetic particles like iron dust from a syringe needle) cause broadening of the absorption peaks in the spectrum. Drawing the solution into a syringe through a small wad of cotton removes any solid particles.

When the solution is prepared, fill the NMR tube to a height of about 3 cm with the solution. Your instructor may provide solvent that already contains the proper amount of TMS. If not, add 1–2 drops of TMS to the solution in the NMR tube. Invert the tube several times to mix the TMS with the solution. Some types of caps provided for NMR tubes leak; in this case, hold your finger (protected by a small piece of polyethylene sheeting or plastic kitchen wrap) over the end of the tube while inverting it.

TMS boils at 28° and is therefore quite volatile. If the signal for TMS in your spectrum is not clear, add one more drop to the NMR tube, mix, and rerun the spectrum. If you are using premixed solvent–TMS, keep the bottle tightly capped.

Too much TMS in the sample solution results in a massive signal. If the NMR tube is not spinning at the proper speed, a large TMS signal will also be accompanied by **spinning sidebands** (see below), which can be mistaken for sample signals. If your sample contains an excessive amount of TMS, prepare a fresh solution, add a smaller amount of TMS than before, and rerun the spectrum. Alternatively, set the NMR tube (uncapped) in the fume hood and allow some of the TMS to evaporate.

B. Instrumental Spectrum Features

Ringing and spinning sidebands   Two patterns that frequently appear in NMR spectra are *ringing* and *spinning sidebands* (see Figure 16.4). These patterns do not arise from the sample itself.

Ringing in a spectrum is a sign of a properly tuned instrument and always appears on only one side of a peak—the side to which the pen is traveling. (The right-hand side if the pen is traveling from left to right on the spectrum paper.)

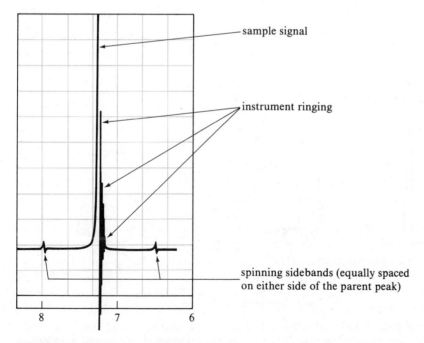

FIGURE 16.4   A proton NMR singlet showing the signal peak (a singlet), instrument ringing, and spinning sidebands.

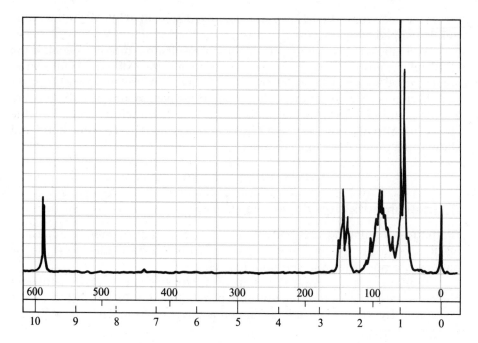

low field strength (60 MHz)

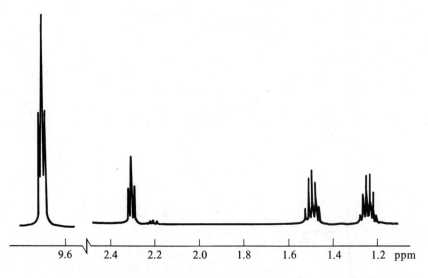

high field strength (500 MHz)

FIGURE 16.5 NMR spectra of *n*-pentanal at low and high field strengths. (From *Theory and Practice in the Organic Lab* by Landgrebe. Copyright © 1993 by Wadsworth, Inc. Reprinted by permission.)

Spinning sidebands appear as pairs of peaks (generally more than one pair) equally spaced on either side of the signal of the parent peak. The position of the spinning sidebands relative to the signal peak is a function of the speed at which the sample is spun in the magnetic field. The position of the spinning sidebands will change when the spin rate is changed. The positions of ringing and the sample signal are independent of spin rate.

High field strength   Modern FT–NMR (Fourier transform) spectrometers routinely use 200, 300, and 500 MHz for obtaining proton spectra. Higher fields not only provide better resolution of proton signals that might overlap at lower fields but also can sometimes result in simplification or multiplets. As the field strength increases, chemical shift differences increase but coupling constants remain fixed. Thus, a higher-order multiplet at a low field can begin to approach a first-order multiplet at a high field. For example, at low field strength (60 MHz), the NMR spectrum of *n*-pentanal shows only ill-defined multiplets. However, at high field strength (500 MHz), the spectrum shows well-defined multiplets that can be easily interpreted (see Figure 16.5).

C. Using Deuterium in Proton NMR Spectroscopy

Any nucleus with spin can absorb radio waves when it is subjected to a magnetic field. However, each type of nucleus has its own unique combination of magnetic field strength and radio frequency required for resonance. The average resonance frequency of a proton in a magnetic field of 14,092 gauss (G) is 60 MHz. By contrast, the resonance frequency for deuterium ($_1^2H$) at 14,092 G is only 2.3 MHz. Therefore, in an NMR spectrum obtained at 60 MHz, normal protons ($_1^1H$) absorb energy, but deuterium nuclei do not.

The fact that deuterium does not absorb energy in proton NMR spectra has several important consequences. NMR solvents are often deuterated so that they do not interfere with proton NMR spectra. Chloroform-d (deuteriochloroform, $CDCl_3$) is a common solvent for NMR work.

Another use of deuterium in NMR spectra is the substitution of a deuterium atom for a hydrogen atom in a compound to simplify an otherwise complex spectrum or to identify a particular proton. For example, the substitution of deuterium can be obtained to identify a peak arising from OH or NH protons. The sample is shaken with a few drops of D_2O, the NH or OH protons undergo chemical exchange, and the peak in the NMR spectrum that changes is thus identified as the OH or NH peak. In this case, a new peak will appear at a δ value of about 5 ppm because of the formation of DOH.

$$ROH + D_2O \rightleftharpoons ROD + DOH$$

old absorption   new absorption

D. Shift Reagents

There are occasions when absorptions from various protons in a molecule overlap to such an extent that little useful information can be obtained from the spectrum. If the spectrum has been obtained at a comparatively low field, one solution would be to run it on an instrument with a much higher field, with the hope that the absorptions will be better resolved.

Another approach to improving resolution is to add small amounts of lanthanide complexes, called **lanthanide shift reagents**, to the sample. Dramatic increases in the chemical shift differences between protons are often observed.

The lanthanide elements, atomic numbers 58–71, have the $4f$ energy level filled with electrons. The common oxidation state for these elements is $+3$, and most of these ions have an unpaired electron. Consequently, they are paramagnetic. A common reagent used as a shift reagent is tris(6,6,7,7,8,8-heptafluoro-2,2-dimethyl-3,5-octanedionato)europium(III), abbreviated $Eu(fod)_3$ (also known as Siever's reagent). The europium in this shift reagent complexes with the oxygen or nitrogen of alcohols, ketones, ethers, amines, and esters. The paramagnetic effect of the europium atom causes protons closer to it in the complex to undergo a larger change in chemical shift than those protons located farther away. The effect of adding $Eu(fod)_3$ to 1-butanol is shown Figure 16.6.

$Eu(fod)_3$
Siever's reagent, a lanthanide shift reagent

Chiral shift reagents   The utility of shift reagents can be extended by using optically active reagents, such as tris(trifluoroacetyl-*d*-camphorato)praseodymium(III), abbreviated $Pr(facam)_3$.

$Pr(facam)_3$
a chiral lanthanide shift reagent

Unlike europium, which generally shifts nearby protons downfield, praseodymium usually shifts nearby protons upfield. Figure 16.7 shows

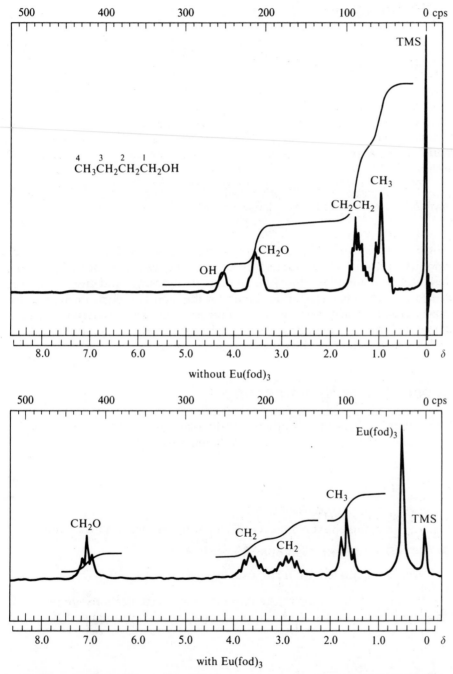

FIGURE 16.6 NMR spectra (60 MHz) of 1-butanol with and without Eu(fod)$_3$ shift reagent. (From *Theory and Practice in the Organic Lab* by Landgrebe. Copyright © 1993 by Wadsworth, Inc. Reprinted by permission.)

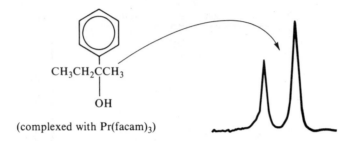

CH₃CH₂CCH₃

OH

(complexed with Pr(facam)₃)

FIGURE 16.7 A portion of the upfield NMR spectrum of partially resolved 2-phenyl-2-butanol with Pr(facam)₃, showing the singlets of the two enantiomeric methyl groups. (From *Theory and Practice in the Organic Lab* by Landgrebe. Copyright © 1993 by Wadsworth, Inc. Reprinted by permission.)

the upfield region of the spectrum of partially resolved 2-phenyl-2-butanol [enriched in the (R)-enantiomer]. Note that the two members of the di-astereoisomeric complex, the (R)-enantiomer-Pr(facam)₃ and the (S)-enanti-omer-Pr(facam)₃, give a separate singlet for the methyl group attached at the benzylic position. Furthermore, the relative areas under these two peaks give a direct measure of the optical purity of the sample.

16.3   Ultraviolet and Visible Spectroscopy

The relationship between absorbance and molar absorptivity of a band in the UV/visible range is calculated by the following formula:

$$\varepsilon = \frac{A}{Cl}$$

where A = observed absorbance

ε = molar extinction coefficient at the absorption maximum

C = concentration in moles per liter (M)

l = path length in cm, usually 1 cm

For calculation of the extinction coefficient, exact concentrations are required. Consequently, ultraviolet and visible spectra are always obtained by using solutions.

A. Cells

Solvent cells, called *cuvettes*, are used to hold the sample solution (see Figure 16.8). Cuvettes hold only a few milliliters of solution. If the wavelength of the

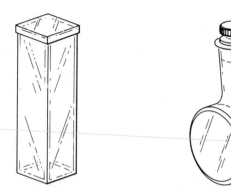

FIGURE 16.8 Cells, or cuvettes. (From *Theory and Practice in the Organic Lab* by Landgrebe. Copyright © 1993 by Wadsworth, Inc. Reprinted by permission.)

spectrum is above 300 nm, then Pyrex® cuvettes can be used. However, because Pyrex® itself absorbs almost all light below 300 nm, obtaining a spectrum at wavelengths lower than 300 nm requires quartz cells.

B. Solvents

Solvents most commonly used for ultraviolet work include alcohols, such as methanol, and hydrocarbons, such as *n*-hexane, cyclohexane, and isooctane. Acetonitrile is a useful aprotic polar solvent. Commercial solvents suitable for use in UV spectroscopy are frequently labeled "spectro-grade." These are not

TABLE 16.1 Ultraviolet solvents

Solvent	Lower wavelength limit[a] (nm)
water	205
ethanol (95% or absolute)	210
hexane	210
cyclohexane	210
methanol	210
diethyl ether	210
acetonitrile	210
tetrahydrofuran	220
dichloromethane	235
chloroform	245
carbon tetrachloride	265
benzene	280

[a]The wavelength at which a cell with a 1-cm path length has an absorbance of approximately 1.

necessarily reagent-grade chemicals, but have been specially treated to remove impurities absorbing in the ultraviolet region.

Table 16.1 lists the common solvents used in ultraviolet and visible spectroscopy, along with their lower wavelength limit.

The position of the absorption bonds are solvent-dependent. Therefore, the solvent must be reported along with the wavelengths of the peaks.

C. Concentration

Instruments have their greatest accuracy when the absorbance is between 0.2 and 0.7 absorbance units. Therefore, the concentration of the sample must be adjusted to bring the absorbance within this range.

Suggested Readings
Infrared Spectroscopy

Bellamy, L. J. *The Infrared Spectra of Complex Organic Molecules*. 3rd ed. New York: Wiley, 1975.

Conley, R. T. *Infrared Spectroscopy*. 2nd ed. Boston: Allyn & Bacon, 1972.

Dyer, J. R. *Application of Absorption Spectroscopy of Organic Compounds*. Englewood Cliffs, N. J.: Prentice-Hall, 1965.

Nakanishi, K., and Solomon, P. H. *Infrared Absorption Spectroscopy*. San Francisco: Holden-Day, 1977.

Silverstein, R. M., Bassler, C. G., and Morrill, T. C. *Spectrometric Identification of Organic Compounds*. 5th ed. New York: Wiley, 1990.

Pouchert, C. J., ed. *The Aldrich Library of Infrared Spectra*. 3rd ed. Milwaukee: Aldrich Chemical Co., 1981

Sadtler Standard Spectra. Philadelphia: Sadtler Research Laboratories, 1978.

Pouchart, C. J., ed. *The Aldrich Library of FT-IR Spectra*. Milwaukee: Aldrich Chemical Co., 1989.

NMR Spectroscopy

Dyer, J. R. *Applications of Absorption Spectroscopy of Organic Compounds*. Englewood Cliffs, N. J.: Prentice-Hall, 1965.

Jackman, L. M., and Sternhell, S. *Applications of Nuclear Magnetic Resonance in Organic Chemistry*. 2nd ed. New York: Pergamon Press, 1969.

Mathieson, D. W. *Nuclear Magnetic Resonance for Organic Chemists*. New York: Academic Press, 1967.

Paudler, W. W. *Nuclear Magnetic Resonance*. Boston: Allyn & Bacon, 1971.

Silverstein, R. M., Bassler, G. C., and Morrill, T. C. *Spectrometric Identification of Organic Compounds*. 5th ed. New York: Wiley, 1990.

Ault, A., and Dudek, G. *An Introduction to Proton NMR Spectroscopy*. San Francisco: Holden-Day, 1976.

Becker, E. D. *High Resolution NMR: Theory and Chemical Applications*. 2nd ed. New York: Academic Press, 1980. (For the more advanced student.)

Shapiro, R. H., and Depuy, C. H., eds. *Exercises in Organic Spectroscopy*. 2nd ed. New York: Holt, Rinehart & Winston, 1977.

Pouchert, C. J., and Campbell, J. R., eds. *The Aldrich Library of NMR Specra*. Milwaukee: Aldrich Chemical Co., 1983.

Macomber, R. S. *NMR Spectroscopy—Basic Principles and Applications*. San Diego: Harcourt Brace Jovanovich, 1988.

Duddeck, H., and Dietrich, W. *Structure Elucidation by Modern NMR—A Workbook*. New York: Springer-Verlag, 1989.

Sadtler Standard Spectra. Philadelphia: Sadtler Research Laboratories, 1978.

Ultraviolet Spectroscopy

Silverstein, R. M., Bassler, G. C., and Morrill, T. G. *Spectrometric Identification of Organic Compounds*. 5th ed. New York: Wiley, 1990.

Dyer, J. R. *Application of Absorption Spectroscopy of Organic Compounds*. Englewood Cliffs, N. J.: Prentice-Hall, 1965.

Faffe, H. H., and Orchin, M. *Theory and Applications of Ultraviolet Spectroscopy*. New York: Wiley, 1962.

Problems

16.1 A student attempts to run a spectrum of a mineral oil mull. The student adjusts the baseline as high as it will go but cannot get it above 40% T. Suggest at least two possible reasons for the student's problem.

16.2 A student carries out an infrared spectrum of a solid compound in CCl_4 solution. The student is surprised when the pen leaves the baseline at 85% T and rises to the *top* of the paper between 725–800 cm^{-1} (12–14 μm), and then returns to the baseline. What caused this "reverse" peak?

16.3 The infrared spectrum (thin film) of a compound with the molecular formula C_7H_5N shows weak absorption at 3100 cm^{-1} (3.2 μm), moderately strong absorption at 2230 cm^{-1} (4.5 μm), and three peaks between 1400 and 1500 cm^{-1} (6.7–7.1 μm) as the principal absorption. Suggest a structure for this compound.

16.4 Sodium 3-(trimethylsilyl)propanesulfonate, $(CH_3)_3SiCH_2CH_2CH_2SO_3^-Na^+$, is a commonly used internal standard for NMR spectra of D_2O solutions. What would be the advantages and disadvantages of using this compound as an internal standard?

16.5 A student is running the NMR spectrum of a D_2O solution of a compound using acetone as an internal standard. How can the student relate the δ values of his spectrum to TMS?

Introduction to the Chemical Literature

17.1 The Chemical Literature

The complete, written, published record of chemical knowledge is referred to as the **chemical literature.** The **primary literature,** or **original literature,** comprises the original reports of compound preparation, compound characterization, mechanistic studies, and so forth. These reports usually appear in research journals (as articles, notes, and communications) and in patent disclosures. Table 17.1 lists just a few of the journals available.

From the primary literature, information flows into the **secondary literature.** The secondary literature consists of compilations of data; articles reviewing entire areas of research; textbooks; abstracts, or summaries, of individual current research articles; and other such publications.

While the primary literature concerns itself with the reports of new findings, the goals of most secondary literature publications are to *summarize and correlate chemical knowledge*. This summarization and correlation are necessary because of the sheer quantity of chemical knowledge. There are approximately 4 million known compounds, and the number of new compounds recorded yearly by the Chemical Abstracts Service of the American Chemical Society is currently about $\frac{1}{3}$ million! Also, as new facts are discovered, chemical theories undergo modification, and previous errors in the literature are corrected. No person could possibly keep up with current chemical news and views by reading the primary literature alone. The secondary literature allows us to keep abreast of an area of study; to find the physical constants of a compound (which might have been reported erroneously in one journal in 1947 and reported correctly in another journal in 1952); or to find a pertinent original journal article. Figure 17.1 depicts the flow of information from the research laboratory through the literature.

TABLE 17.1 A few chemical research periodicals

Name	Abbreviation[a]	Language
Journal of the American Chemical Society	*J. Am. Chem. Soc.*	English
Journal of Organic Chemistry	*J. Org. Chem.*	English
Canadian Journal of Chemistry	*Can. J. Chem*	English
Journal of the Chemical Society	*J. Chem. Soc.*[b]	English
Bulletin of the Chemical Society of Japan	*Bull. Chem. Soc. Jpn.*	English, French, German

[a]These are the abbreviations used in the *Journal of the American Chemical Society*; other publications may use different abbreviations.
[b]In 1965, the *Journal of the Chemical Society* was becoming sufficiently lengthy that it was divided into three parts: *A, B,* and *C.* In 1972, *C* (physical organic chemistry) was renamed *Perkin Transactions I,* and *B* (organic and bio-organic) was renamed *Perkin Transactions II.* (Other section names for the other areas of chemistry are also encountered.) In references to the post-1972 journals, these names are abbreviated and arabic numerals are used: for example, *J. Chem. Soc., Perkin Trans. 1.*

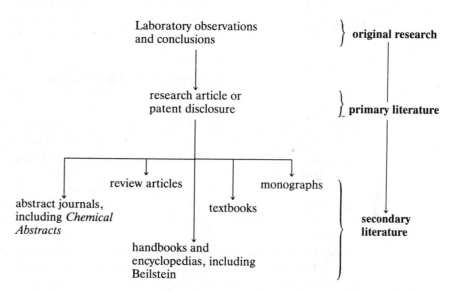

FIGURE 17.1 Flow of information from the laboratory into the chemical literature.

· · · · · · · · · ·
17.2 Retrieving Information from the Literature

How we retrieve information from the chemical literature depends on the type of information desired. The ultimate goal of information retrieval would be to find the original research reports that contain the raw data from which all other reports are derived. While this is a worthy goal, it is often neither necessary nor practical. If, for example, you need only the melting point of an organic compound, you can stop your literature search once you have found the value in a reference work. Or if you want to survey a general topic of organic chemistry (such as the Grignard reaction), you would probably look for review articles or book chapters. (It might take several years to read every original article in which Grignard reactions are mentioned!) On the other hand, if you wish to repeat a particular synthetic scheme already reported in the literature, you would want to study the original journal article to find out the exact reaction conditions used and what problems were encountered.

Another aspect of information retrieval is the year that the information was published. Even with on-line computer retrieval systems and special publications, recent information is the most difficult to find because it has not yet been included in the secondary literature.

In the following sections, we outline the various techniques of searching the secondary literature and then briefly introduce the use of Beilstein's *Handbuch der Organischen Chemie* and *Chemical Abstracts*. More comprehensive instructions for searching the literature are found in the following references.

R. F. Gould, ed. *Searching the Chemical Literature*. Advances in Chemistry Series No. 30. Washington, D. C.: American Chemical Society, 1961.

H. M. Woodburn. *Using the Chemical Literature. A Practical Guide*. New York: Dekker, 1974.

R. E. Maizell. *How to Find Chemical Information*. New York: Wiley, 1979.

R. T. Bottle, ed. *Use of Chemical Literature*. 3rd ed. London: Butterworths, 1979.

A. Somerville, "Information Sources for Organic Chemistry." *J. Chem. Ed.* **1991,** *68(7)*, 552.

The following references provide an introduction to on-line searching and the associated computer protocol.

H. M. Dess, M. Kesselman, and G. M. Muha. "Introducing On-Line Searching of *Chemical Abstracts* in the Undergraduate Curriculum." *J. Chem. Ed.* **1990,** *67(11)*, 947.

J. M. Miller. "Independent Student Searching of the Chemical Abstracts Files." *J. Chem. Ed.* **1989,** *66(1)* 24.

M. Krumpolc, D. Trimakas, and C. Miller. "Searching Chemical Abstracts Online in Undergraduate Chemistry, Part 1." *J. Chem. Ed.* **1987,** *64(1)*, 55.

M. Krumpolc, D. Trimakas, and C. Miller. "Searching Chemical Abstracts Online in Undergraduate Chemistry, Part 2." *J. Chem. Ed.* **1989,** *66(1)*, 28.

C. Carr. "Aids for Teaching Online Searching of the Chemical Literature." *J. Chem. Ed.* **1989,** *66(1)*, 21.

17.3   How to Find General Information

Finding general information in the literature, such as information concerning the reduction of aldehydes, is often more challenging than finding a specific fact. The reason is that general terms such as *reduction* and *aldehydes* may take up a considerable amount of space in a comprehensive index (or they might not appear at all).

The best approach to finding information about a general topic is to start with a fairly general reference book. For example, a textbook may include a discussion with references to review articles. Some other publications that are useful for finding general information follow.

S. Coffey, ed. *Rodd's Chemistry of Carbon Compounds*. 2nd ed. New York: Elsevier, 1964.

Organic Syntheses. New York: Wiley, 1920–present. Tested synthetic procedures for specific compounds.

Organic Reactions. New York: Wiley, 1942–present. Detailed discussions of important reaction types (such as Friedel-Crafts alkylations) from the laboratory point of view.

A. Weissberger, ed. *Technique of Organic Chemistry*. New York: Interscience Publishers, 1949–present.

L. F. Fieser and M. Fieser. *Reagents for Organic Synthesis*, Vols. 1–9. New York: Wiley, 1968–present.

There are many other books, monographs on specific topics, and review journals and books (such as *Chemical Reviews*, *Accounts of Chemical Research*, and *Advances in Drug Research*) that review and summarize various fields. Check the library catalog.

The next step in searching for general information is to consult the subject indexes of *Chemical Abstracts*. (See the discussion on using *Chemical Abstracts* in Section 17.6.)

17.4   How to Find Information about a Specific Organic Compound

There are numerous ways to find information about a particular compound. If you are simply looking for physical constants of a compound, first check the extensive tables in the handbooks. If you cannot find the name of the compound in the tables, use the molecular formula index—the compound may be listed under a different name.

Handbooks

CRC Handbook of Chemistry and Physics. Boca Raton, Fla.: CRC Press, Inc., published anually or biennially.

Lange's Handbook of Chemistry. New York: McGraw-Hill, many editions.
Z. Rappoport, ed. *Handbook of Tables for Organic Compound Identification.* 3rd ed. Boca Raton, Fla.: CRC Press, Inc., 1967.
Merck Index. 10th ed. Rahway, N.J.: Merck and Co., 1983.
Heilbron's Dictionary of Organic Compounds. 5th ed., Vols. 1–5. New York: Oxford University Press, 1982.

Spectral compilations

Proton Nuclear Magnetic Resonance Spectra. Philadelphia: Sadtler Research Laboratories, continuously updated.
Sadtler Standard Spectra. Philadelphia: Sadtler Research Laboratories, continuously updated: infrared spectra.
C. J. Pouchert, ed. *The Aldrich Library of Infrared Spectra.* 3rd ed. Milwaukee: Aldrich Chemical Co., Inc., 1981.
C. J. Pouchert and J. R. Campbell, eds. *The Aldrich Library of NMR Spectra.* Milwaukee: Aldrich Chemical Co., Inc., 1974.

If the desired information cannot be gleaned from a handbook, then you must search for an original research citation. If you have access to a computer on-line search program, this would be your first choice for a search tool. However, on-line searches are expensive unless you are skilled at querying the system. Nevertheless, for a research question, an on-line search is the method of choice because it is up to date and very fast. A less expensive, but considerably more time-consuming, method is a hand search through Beilstein, followed by a search of the indexes of *Chemical Abstracts*

• • • • • • • • • •
17.5 Beilstein's *Handbuch der Organischen Chemie*
A. Background

Beilstein's *Handbuch der Organischen Chemie* (Berlin: Springer-Verlag, 1918–present), commonly referred to as simply "Beilstein," is a monumental compilation of organic data. Beilstein was first published by Friedrich Karl Beilstein in 1881–1882. After the third edition, the German Chemical Society acquired the rights to this work and published the fourth edition, which is the edition used today. Although it is written in German, the organic formulas, numbers, etc., are the same as those in books written in English. For this reason, a minimum of German vocabulary enables the beginning student to find useful data. (Table 17.2 lists some abbreviations used in Beilstein.)

The fourth edition of Beilstein consists of a main series, called **das Hauptwerk,** of 27 volumes, plus supplemental series, or **Erganzungswerk,** each of which also has 27 volumes. The main series contains data on *all known organic compounds through 1909,* including all known syntheses, reac-

TABLE 17.2   Some abbreviations used in Beilstein[a]

Abbreviation	German term	English equivalent
Kp	Siedepunkt	boiling point
D	Dichte	density
E	Erstarrungspunkt	freezing point
F	Schmelzpunkt	melting point
n	Brechungsindex	refractive index
Zer	Zersetzung	decomposition

[a]Examples:
F : 20° (reference): a melting point of 20° followed by the original reference where the value was reported.
Kp_{760} 240° (korr): a corrected boiling point of 204° at a pressure of 760 mm Hg.

tions, physical constants, and references to the original literature. Supplement I extends the coverage through 1919 and includes new data for compounds appearing in the main series as well as data for compounds discovered between 1910 and 1919. Later supplements have been published in whole or in part, as the volumes are completed. Table 17.3 lists the main series and first four supplements, their German names, and years of coverage. Because of the time lag between an original publication and its inclusion in Beilstein, this reference work is most useful for searching the older literature, up to about 1940.

B. General Organization

The entries in Beilstein are arranged according to structure rather than by name. While a complete understanding of Beilstein's organizational system is not necessary for the successful use of this reference work, a general idea of the overall organization is helpful. The main series and each supplemental series have the *same organization*. Methane, for example, is the first compound listed in Volume I of the main series and is also the first compound listed in Volume I of each of the supplements. Therefore, once an entry has been located in the main series, it is an easy task to locate the same entry in the supplements.

TABLE 17.3   Series in Beilstein's *Handbuch der Organischen Chemie*

Series and supplements	German name	Coverage
Main Series	Hauptwerk	up to 1910
Supplement I	Erstes Erganzungswerk (E I)	1910–1919
Supplement II	Zweites Erganzungswerk (E II)	1920–1929
Supplement III	Drittes Erganzungswerk (E III)	1930–1949
Supplement IV	Viertes Erganzungswerk (E IV)	1950–1959

The first four volumes in the main series (and also in the supplements) cover open-chain (acyclic) compounds; Volumes 5–16 cover compounds containing carbon rings (carbocyclic compounds); and Volumes 17–27 cover heterocyclic compounds. Within each of these three major classifications, compounds are entered in the following order: *hydrocarbons, hydroxy compounds, carbonyl compounds,* and *carboxylic acids,* followed by other compound classes that we will not discuss.

Open-chain compounds (Volumes 1 – 4):

$$RH \longrightarrow ROH \longrightarrow RCHO \quad \text{and} \quad R_2C{=}O \longrightarrow RCO_2H, \quad \text{etc.}$$

Carbocyclic compounds (Volumes 5 – 16):

$$RH \longrightarrow ROH \quad \text{and} \quad ArOH \longrightarrow RCHO \quad \text{and} \quad R_2C{=}O \longrightarrow RCO_2H, \quad \text{etc.}$$

Heterocyclic compounds (volumes 17 – 27):

$$\text{parent heterocycle} \longrightarrow ROH \quad \text{and} \quad ArOH \longrightarrow RCHO$$

$$\text{and} \quad R_2C{=}O \longrightarrow RCO_2H, \quad \text{etc.}$$

Within each of the functional group classes, the simplest structure (lowest molecular weight and highest degree of saturation, such as methane for the open-chain hydrocarbons or methanol for the hydroxy compounds) is listed first. The "derivatives" of each main entry follow that entry; for example, cyclohexyl methyl ether is found following cyclohexanol.

While it is quite possible to locate an entry in Beilstein using only the Beilstein system, a considerable amount of time must be spent learning the system. For those who wish more information, the following references should be consulted.

How to Use Beilstein. Frankfurt/Main: The Beilstein Institute; distributed by Springer-Verlag, New York, Inc., 175 Fifth Ave., New York, N. Y. 10010.

O. Weissback. *The Beilstein Guide: A Manual for the Use of Beilstein's Handbuch der Organischen Chemie.* New York: Springer-Verlag, 1976.

J. Sunkel, E. Hoffmann, and R. Luckenbach. "Straightforward Procedure for Locating Chemical Compounds in the Beilstein Handbook." *J. Chem. Ed.* **1981,** *58,* 982.

C. How to Locate a Compound in Beilstein

The easiest way to find a compound in Beilstein is to find the Beilstein reference in a handbook (such as Lange's *Handbook of Chemistry* or the *Handbook of Chemistry and Physics*) or to use one of the Beilstein indexes. Each volume of Beilstein contains indexes of compound names and molecular formulas. In addition, Volumes 28 and 29 of the main series and each supplemental series contain *cumulative indexes.*.

Find the molecular formula in Volume 29 (*General Formelregister*) of Supplement II or other series.

↓

Find the correct entry from (a) the German name, or (b) checking the structural formulas in the various entries.

↓

Use the page references listed in the index for the main series and each supplement and/or use the cross-references at the top of the pages to find a compound.

FIGURE 17.2 A summary of how to find information about a compound in Beilstein's *Handbuch der Organischen Chemie*.

Because a knowledge of German organic nomenclature is needed to locate a compound using the cumulative name index (Volume 28, *General Sachregister*), we will describe the use of the molecular formula index (Volume 29, *General Formelregister*). Volume 29 of Supplement II is contained in three books and lists all the compounds found in the main series as well as those in the first two supplements. Compounds are listed by increasing carbon content (all the C_1 compounds, followed by the C_2 compounds, etc.). The use of the molecular formula index and the key notations therein are best explained with the aid of a specific example. The general procedure used to locate a compound in Beilstein is summarized in Figure 17.2.

EXAMPLE Locate cyclohexanol in the main series and in the first *three* supplements of Beilstein.

1) Write the structural formula for cyclohexanol and determine the molecular formula.

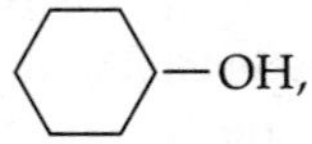

⬡—OH, $C_6H_{12}O$

2) Find the $C_6H_{12}O$ entries in the formula index (Volume 29) of Supplement II or other series.

These entries appear on page 221 in the first bound book of Volume 29 of Supplement II.

3) Inspect the German listing to find the name that corresponds most closely to the English name.

Our example is simple—the German and the English names are identical: cyclohexanol. In more difficult cases, (a) use a German–English chemical dictionary, or (b) look up each entry and check the structural formulas.

4) Write down the Beilstein page references immediately following cyclohexanol.

> Cyclohexanol *6*, 5, I4, II5

These references are interpreted as follows.

6, 5: Cyclohexanol is found in Volume 6 (*Band* 6), page 5, of the main series.
I4: This compound is found in Volume 6, page 4, of Supplement I (E I).
II5: The compound is found in Volume 6, page 5, of Supplement II (E II).

5) Find cyclohexanol in the main series.
In Volume 6, page 5, you will find:

> 3. *Oxy-Verbindungen* $C_6H_{12}O$.
> 1. Cyclohexanol, Hexahydrophenol $C_6H_{12}O$ = ...

Note that the structural formula is given immediately following the molecular formula. The information concerning cyclohexanol follows the heading. Each informational entry is referenced to the original literature. (The journal abbreviations are listed in the front of each book.)

6) Find cyclohexanol in Volume 6 of Supplements I and II.
Cyclohexanol is found on page 4 in Volume 6 of Supplement I. Note that the format is identical to that in the main series.
Cyclohexanol is found on page 5 in Volume 6 of Supplement II. Again, note that the format is identical.

7) Find cyclohexanol in the supplements published after Supplement II.
All supplements of Beilstein are cross-referenced to the main series; therefore, you can find a compound in the newer supplements without having to use another index.
At the top of page 4 in Volume 6 of Supplement I and page 5 in Volume 6 of Supplement II, you will find the following notations:

> VI, 4–5   (in Supplement I)

> H 6, 5–6   (in Supplement II

These notations refer to the pages in Volume 6 of the main series (H, or *Hauptwerk*) on which the same compounds (including cyclohexanol) are found.

To find cyclohexanol in Volume 6 of Supplement III, you need only turn to the page that has "H 6, 5" centered at the top (page 10 of Volume 6, of Supplement III).   ●

. . . .
17.6   *Chemical Abstracts*

A. Background

The most comprehensive collection of modern chemical information is *Chemical Abstracts* (*Chem. Abstr.*, *CA*). *Chemical Abstracts* have been published continuously since 1907. Today, the Chemical Abstracts Service abstracts hundreds of thousands of documents per year from about 14,000 periodicals. The collective index (not the abstracts themselves) just for the five-year period 1977–1981 occupies over 131,455 pages!

The organization and goals of *Chemical Abstracts* are different from those of Beilstein. Unlike Beilstein, which covers only organic compounds (in ten-year units), *Chemical Abstracts* cover all areas of chemistry and attempt to publish concise abstracts of every article and patent as soon as possible (usually three to four months from the original publication data). These abstracts are organized within each printed volume into general fields (organic chemistry, biochemistry, and so on).

The format of *Chemical Abstracts* is in a state of continuous change. Originally, one volume (consisting of several bound books) was published corresponding to one year of the scientific literature. As the number of scientific publications increased, this became impractical; and beginning in 1962 (Volume 56), each volume has covered only six months of the primary literature.

The manner in which index references are given in *Chemical Abstracts* has also changed over the years. In the early years, the reference to an abstract was by volume and page number. As the pages became more crowded, columns, instead of pages, were numbered and superscript numbers (and later, letters) were added to the index references to denote the position on the page or column. Starting in 1967 (Volume 66), each abstract has been given its own number. Fortunately, these changes create no confusion for someone using *Chemical Abstracts*; the meaning of a reference becomes self-evident from the appearance of a page in a particular volume.

To a user of *Chemical Abstracts*, a far more important change is that of nomenclature, which has undergone major revisions over the decades. The older issues emphasized trivial names. The newer issues emphasize a modified IUPAC system, called the "Chemical Abstracts system." Today, very little trivial nomenclature is found in *Chemical Abstracts*. For example, toluene is now indexed as "benzene, methyl-".

B. Indexes

Each volume of *Chemical Abstracts* has an author index, a subject index, and a molecular formula index covering the issues of that six-month (or one-year) period. Until 1956, *collective indexes* were published every ten years; however, from 1957 to the present, these have been published every five years. (These collective indexes are not cumulative, as are the Beilstein indexes.)

Because of the changes in the indexing system over the years (due, in large part, to computerization), the techniques for finding information before 1966 and after 1966 are somewhat different (see Section 17.6D).

C. Nomenclature

In the subject indexes of *Chemical Abstracts*, compounds are generally listed in the alphabetical order of their parents ("1-pentanol, 3-methyl" instead of "3-methyl-1-pentanol"). As we have mentioned, some of the parent index names have changed over the years. Fortunately, *Chemical Abstracts* has published guides to finding compounds in the subject indexes.

For the earlier volumes: See the nomenclature guide in the introduction to the subject index of Volume 56 (January–June 1962).

For the later volumes: See the *Index Guides* for the 8th and 9th Collective Indexes. (These guides contain a selection of index names, an alphabetical listing of cross-references, synonyms, and so on.)

D. How to Find a Compound in *Chemical Abstracts*

Pre-1966 information The 7th Collective Index, covering Volumes 56–65 (1962–1966), was the last of the old indexes. To find information published prior to 1966, start with this Collective Index and search backward in time, using the 6th Collective Index next, and so forth. Both the subject indexes and the formula indexes will usually have to be consulted.

EXAMPLE Search the 7th Collective Index of *Chemical Abstracts* for abstracts dealing with the kinetics of the acetylation of cyclohexanol.

1) Write the formula and determine the molecular formula of the compound.

$$\bigcirc\!\!-\text{OH}, \quad C_6H_{12}O$$

2) Find the $C_6H_{12}O$ listings in the formula index.

The listings start on page 511F of the first volume of the 7th Collective Formula index. You will find that there are no listings for cyclohexanol; all are contained in the subject index. (This fact is stated in the parenthetical statement immediately following the start of $C_6H_{12}O$ headings on page 511F, beginning with "See also . . .")

3) Find the compound in the subject index.

The listings for cyclohexanol begin on page 69475 and continue to page 69495.

4) Look for the specific subject.

Acetylation of cyclohexanol is the second major entry under cyclohexanol on page 69475.

Acetylation of, *59*: 1513e
   catalysts for, *60*: 14505g
   kinetics of, *65*: 7257g, 8727h
   micro, *59*: 8632h

There are two references for our subject: Volume 65, column 7257, position g; and Volume 65, column 8727, position h.

5) Find the specific abstract. (We will consider only the second abstract here.)

At the beginning of the *65*: 8727h abstract you will find:

Conformational analysis, XII. Acetylation rates of substituted cyclohexanols. The kinetic method of conformational analysis. Ernest L. Eliel and Francis J. Biros (Univ. of Notre Dame, Notre Dame, Indiana). *J. Am. Chem. Soc.* **88**(14), 3334–43 (1966) (Eng); . . . .

Post-1966 information The use of the indexes of *Chemical Abstracts* starting with the 8th Collective Index has been considerably simplified by the *Index Guide* and registry numbers. The *Index Guide* provides a cross-reference to the indexes: by first consulting the *Index Guide*, you can determine the word or compound to look up in the subject index.

The **registry number** is a unique computer-generated number that is assigned to each chemical substance to provide identification. The number for the cyclohexanol is [108-90-0]. With the registry number, a chemist can consult the Chemical Abstracts Service *Registry Handbook* to find the index name and molecular formula. The first *Registry Handbook* was published in

1974 and covered the literature from 1965 to 1971. Supplements have been issued since then.

The general procedure for using the more recent indexes of *Chemical Abstracts* follows.

1) If you wish to find information about a structure, but have only its common (trivial) name:
   (*a*) Look up the name in the *Index Guide* to find the name under which the compound is indexed.
   (*b*) Look up the index name in the *Subject Index* (or the *Chemical Substance Index* if you are using the 8th or later Collective Index).
   (*c*) Find the abstract(s).
2) If you have a molecular formula of the substance:
   (*a*) Use the *Formula Index* to find both the indexed name and specific references.
   (*b*) Use the *Subject Index* (or the *Chemical Substance Index*) to obtain additional references.
   (*c*) Find the abstract(s).
3) To search for information about a general subject:
   (*a*) Use the *Index Guide* to find out how the subject is indexed.
   (*b*) Go to the *Subject Index* (or the *General Subject Index* starting with the 9th Collective Index).
   (*c*) Find the abstract(s).

Figure 17.3 summarizes the rules for finding information in both the earlier and later volumes of *Chemical Abstracts*.

Collective Indexes 8–present

Find the index name in the *Index Guide* or *Formula Index*.

↓

Find abstract references from both the *Subject Index* (or *Chemical Substance Index*) and the *Formula Index*.

Collective Indexes 1–7 (up to 1966)

Find rules for naming in the introduction to the *Subject Index*, Volume 56.

↓

Find abstract references from both the *Subject Index* and the *Formula Index*.

FIGURE 17.3   Summary of how to find information in *Chemical Abstracts*.

Problems

17.1   Using the *Handbook of Chemistry and Physics*, find the melting points of the following compounds.
(a) *trans*-1,2-Cyclohexanedicarboxylic acid
(b) *m*-Aminophenol
(c) 2,5-Dibromofuran

17.2   Using *Lange's Handbook of Chemistry*, find the refractive index of each of the following compounds.
(a) Benzyl alcohol
(b) Formamide
(c) 3-Pentenoic acid chloride

17.3   Using the *Merck Index*, find the following information.
(a) The medical use of α-methyl-*p*-tyrosine
(b) The structure and primary literature reference to tocol.
(c) What the Wenker ring closure is

17.4   Use the *Index Guide* to the 9th Collective Index to find the index name for the following.
(a) 1-Cyclohexyl-2-propyn-1-ol
(b) Marvinol 2002

17.5   Use the *Index Guide* to the 8th Collective Index to find the registry numbers for the (E) and (Z) isomers of 13-docosenoic acid.

17.6   Use the 1965–1971 *Registry Handbook* to find the name and molecular formula of the compound with the registry number [1226-05-7].

17.7   Use the 9th Collective Index to find an abstract reference for the antimicrobial spectra of alkylbenzene sulfonic acid esters.

17.8 γ, γ, γ-Trichlorocrotonitrile ($Cl_3CCH{=}CHCN$) has been prepared by the reaction of P_2O_5 and $Cl_3CCH{=}CHCONH_2$. Use the formula index for Volumes 14–40 of *Chemical Abstracts* to find the original article.

17.9   Use Beilstein to find the following information.
(a) The journal in which the density and viscosity of a solution of 2,4-dichloroanimline in isoamyl acetate was published
(b) The boiling point at 7 mm Hg of 1-chloro-1,2-dibromocyclohexane
(c) The melting point of a mercuric chloride–hydrochloride salt of 2-methylhiazole
(d) The solubility of 2-nitro-2-methyl-1-propanol in ethanol and in water

17.10   Use Beilstein to find the journal in which a melting point of 18° for $CH_2{=}CH(CH_2)_9CO_2H$ was reported.

Commonly Used Calculations

This appendix is a brief summary of calculations commonly performed in the laboratory. If you do not understand any one of the concepts presented here, you should review that concept in any good general chemistry text.

In any mathematical calculation, carry along the units of the numbers. In this way, you can determine which units cancel (and which do not cancel) as a check on how you have set up your equation. Also, before mixing any reagents or reporting any percent yield, always look at your calculations and ask yourself, "Is this reasonable?"

1 Molarity

Molarity is defined as *the number of moles of solute in 1.00 liter of solution.* The following equations are used to determine molarity.

$$\text{molarity } (M) = \frac{\text{moles of solute}}{\text{liters of solution}}$$

$$= \frac{\text{g of solute/MW of solute}}{\text{liters of solution}}$$

$$= \frac{\text{g of solute}}{(\text{MW})(\text{liters})}$$

EXAMPLE What is the molarity of an aqueous solution of 5.0 g of NaOH in 100 mL of solution?

$$M = \frac{\text{g}}{(\text{MW})(\text{liters})} = \frac{5.0 \text{ g}}{(40.0 \text{ g/mol})(0.100 \text{ L})} = 1.25 \text{ mol/L}$$

EXAMPLE What weight of solid NaOH would you need to prepare 50 mL of a 2.0M aqueous solution?

$$M = \frac{g}{(MW)(liters)}$$

$$g = (M)(MW)(liters)$$

$$= (2.0 \text{ mol/L})(40.0 \text{ g/mol})(0.050 \text{ L})$$

$$= 4.0 \text{ g} \qquad \bullet$$

2 Normality

The **normality** of a solution is the *number of equivalents of solute in 1.00 liter of solution.* In the organic laboratory, normality is generally encountered only with acids and bases (for example, 6N HCl, 6N H$_2$SO$_4$, or 6N NH$_4$OH). One **equivalent** of acid or base is the weight of the substance that contains 1.00 mole of H$^+$ or OH$^-$.

for a monoprotic acid (such as HCl) or a base containing one OH$^-$ per molecule (such as NaOH):
equivalent weight = molecular weight

Therefore,   normality = molarity

EXAMPLES 6N HCl = 6M HCl
6N NaOH = 6M NaOH

for a diprotic acid, such as H$_2$SO$_4$, or a base containing two OH$^-$ per molecule, such as Ca(OH)$_2$:
equivalent weight = $\frac{1}{2}$ molecular weight

Because each mole contains two equivalents,

normality = 2 × molarity $\bullet$

EXAMPLE What is the molarity of 6N H$_2$SO$_4$?

$$N = 2M$$

$$M = \frac{N}{2} = \frac{6N}{2} = 3 \qquad \bullet$$

EXAMPLE Calculate the molarity of a 2.5N aqueous solution of H$_3$PO$_4$. In this case, one mole of H$_3$PO$_4$ can theoretically supply three moles of H$^+$. The normality is three times the molarity.

$$N = 3M$$

$$M = \frac{N}{3} = \frac{2.5N}{3} = 0.83 \qquad \bullet$$

EXAMPLE What weight of $Ca(OH)_2$ would be necessary to prepare 100 mL of a $0.25N$ solution?

$$N = \frac{\text{no. of equivalents}}{\text{liters}}$$

$$\text{no. of equivalents} = N \times \text{liters}$$

$$= (0.25 \text{ equivalent/L})\,(0.100 \text{ L})$$

$$= 0.025 \text{ equivalent}$$

$$\text{equivalent weight} = \tfrac{1}{2}\text{MW}$$

because $Ca(OH)_2$ can supply two OH^- ions per $Ca(OH)_2$ molecule. Therefore,

$$0.025 \text{ equivalent} = \tfrac{1}{2}(0.025)$$

And since the molecular (formula) weight of $Ca(OH)_2$ is 74.10 g/mol,

$$0.025 \text{ equivalent} = \tfrac{1}{2}\,(0.025 \text{ mol} \times 74.10 \text{ g/mol})$$

$$= 0.93 \text{ g}$$

As a check on N–M conversions, remember that for a given solution N **is always equal to or larger than** M. ●

· · · · · · · · · ·

3 Dilutions

In practice, we are often required to dilute a more concentrated acid (or base) to a less concentrated solution. Because the number of moles or equivalents of acid (or base) is not changed by dilution, the following simple equations allow us to calculate the amount of more concentrated solution needed.

$$M_1V_1 = M_2V_2 \quad \text{or} \quad N_1V_1 = N_2V_2$$

where M_1V_1 *and* N_1V_1 *refer to the concentrated solution, and* M_2V_2 *and* N_2V_2 *refer to the dilute solution*

EXAMPLE What volume of $12M$ HCl is needed to prepare 100 mL of $1.5M$ HCl?

$$M_1V_1 = M_2V_2$$

$$V_1 = \frac{M_2V_2}{M_1}$$

$$= \frac{(1.5M)(0.100 \text{ L})}{12M}$$

$$= 0.0125 \text{ L, or } 12.5 \text{ mL}$$ ●

4 Percent Concentrations

Many common laboratory manipulations require solutions with concentrations reported in percents. These percentages generally refer to **weight-volume percents.** For example, a 5% $NaHCO_3$ solution is an aqueous solution of 5 g of $NaHCO_3$ dissolved in water and then diluted *to* 100 mL (not *with* 100 mL.

$$\text{percent (weight/volume)} = \frac{\text{g of solute}}{100 \text{ mL of solution}}$$

EXAMPLE What weight of NaOH is required to prepare 30 mL of a 15% aqueous solution?

This type of problem may be solved quickly by a simple proportion.

$$\frac{15 \text{ g}}{100 \text{ mL}} = \frac{x \text{ g}}{30 \text{ mL}}$$

$$x = \frac{15 \text{ g}}{100 \text{ mL}} \times 30 \text{ mL} = 4.5 \text{ g}$$

In certain instances, it may be desirable to convert a concentrated solution to a more dilute solution. ●

EXAMPLE What volume of 5.0% $NaHCO_3$ is needed to prepare 7.0 mL of 2.0% $NaHCO_3$?

$$C_1 V_1 = C_2 V_2$$

$$V_1 = \frac{C_2 V_2}{C_1}$$

$$\boxed{\begin{array}{ll} C_1 = 5.0\% & C_2 = 2.0\% \\ V_1 = ? & V_2 = 7.0 \text{ mL} \end{array}}$$

$$V_1 = \frac{2.0\% \times 7.0 \text{ mL}}{5.0\%}$$

$$= 2.8 \text{ mL}$$ ●

5 Percent Yields and Theoretical Yields

A **percent yield** is simply the percent of the theoretical amount of product actually obtained in a reaction.

$$\text{percent yield} = \frac{\text{actual yield}}{\text{theoretical yield}} \times 100$$

EXAMPLE   What is the percent yield when 5.2 g of product is obtained from a theoretical 7.5 g?

$$\text{percent yield} = \frac{5.2}{7.5} \times 100 = 69\%$$

To calculate the theoretical yield, balance the reaction and calculate the moles of reactants. Then calculate the theoretical yield based on the limiting reagent, which is the reagent present in the shortest supply. For example, in the following oxidation of cyclohexanol, 20.0 g of the alcohol is treated with 23.8 g of $Na_2Cr_2O_7 \cdot 2H_2O$ and 26.5 g of H_2SO_4.

1)   Calculate the molecular weights.

2)   Calculate the numbers of moles.

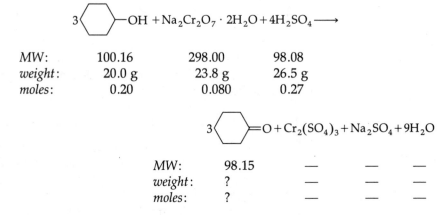

MW:	100.16	298.00	98.08
weight:	20.0 g	23.8 g	26.5 g
moles:	0.20	0.080	0.27

MW:	98.15	—	—	—
weight:	?	—	—	—
moles:	?	—	—	—

3)   Determine the limiting reagent. The required amounts of $Na_2Cr_2O_7 \times 2H_2O$ and H_2SO_4 to oxidize 0.20 mole of cyclohexanol are 0.20/3, or 0.067, mole of the chromate and (4/3) (0.20), or 0.27, mole of H_2SO_4. In this particular reaction, the alcohol and H_2SO_4 are limiting reagents, while the chromate is present in excess.

4)   Determine the theoretical number of moles of product possible. In this case, 0.20 mole of cyclohexanone is the maximum, or theoretical, yield (the same as the number of moles of cyclohexanol, one of the limiting reagents).

5)   Convert the theoretical yield of product to grams.

0.20 mol of cyclohexanone = 0.20 mol × 98.15 g/mol

= 19.6 g (theoretical yield)   ●

Elemental
Analyses

The **weight percents of carbon and hydrogen** in an organic compound are determined by burning a weighed sample of the compound in a special apparatus. The resultant water vapor is collected in a tared chamber containing a drying agent, while the carbon dioxide is trapped in a chamber containing a strong base (which converts the carbon dioxide to the carbonate ion). The chambers are reweighed and the weights of water and carbon dioxide determined by difference. The percent C and percent H in the original compound can then be calculated.

In practice, the problems of ensuring complete combustion and collecting 100% of the gases (uncontaminated by outside moisture or carbon dioxide) are difficult to overcome without the proper equipment.[*] Therefore, most organic chemists do not perform these analyses but send samples to analytical chemists who specialize in this type of analysis. It is, of course, the responsibility of the organic chemist to submit a sample that is as pure as possible. Impurities amounting to 5% of the sample may not measurably affect an infrared or NMR spectrum, but they will invalidate the results of a C and H analysis.

Most organic compounds contain only C, H, and O. A typical analysis report shows only the percent C and percent H. The percent O is usually determined by difference.

If requested to do so, an analytical laboratory can perform analyses for oxygen, the halogens, phosphorus, nitrogen, and other elements. Here, we will discuss compounds containing only C, H, and O.

---

[*]Some of the techniques used in organic analysis are discussed in J. P. Dixon. *Modern Methods in Organic Microanalysis*. Princeton, N. J.: D. Van Nostrand Co., 1968.

• • • • • • • • • • •

1   Determining the Empirical Formula

From the weight percents of the elements in a compound, the **molar ratio** of the elements can be calculated. This is accomplished by dividing each *weight percent* by the *atomic mass* of that element. The following example describes this procedure for a compound containing 38.72% C, 9.72% H, and 51.56% O.

Step 1   Divide percent by atomic mass to determine the molar ratio:

$$\frac{38.72\% \ C}{12.01} = 3.22 \qquad \frac{9.72\% \ H}{1.008} = 9.64 \qquad \frac{51.56\% \ O}{16.00} = 3.22$$

We see that the molar ratio of C, H, and O is $3.22 : 9.64 : 3.22$. These numbers must be converted to small whole numbers by dividing all three values by the smallest value.

Step 2   Divide the values in the ratio by the smallest value:

$$\text{for C:} \ \frac{3.22}{3.22} = 1.00 \qquad \text{for H:} \ \frac{9.64}{3.22} = 2.99 \qquad \text{for O:} \ \frac{3.22}{3.22} = 1.00$$

When the molar ratio has been converted to small whole numbers, the ratio tells us the relative numbers of the atoms in a molecule. In our example, the molar ratio is very close to $1 : 3 : 1$. From these numbers, we can write an **empirical formula.**

Step 3   Write the empirical formula:

CH_3O

Note that this formula is an *empirical* formula, which shows only the *ratios* of the atoms, not their actual numbers in the molecule. The true **molecular formula** might be CH_3O, $C_2H_6O_2$, $C_3H_9O_3$, or any other formula with C, H, and O in a ratio of $1 : 3 : 1$.

EXAMPLE   A compound contains 65.50% C, 9.46% H, and 25.02% O. What is its empirical formula?

Step 1   Divide percent by atomic mass:

$$\frac{65.50\% \ C}{12.01} = 5.45 \qquad \frac{9.46\% \ H}{1.008} = 9.38 \qquad \frac{25.02\% \ O}{16.00} = 1.56$$

Step 2 Divide by the smallest value:

$$C: \frac{5.45}{1.56} = 3.49 \qquad H: \frac{9.38}{1.56} = 6.01 \qquad O: \frac{1.56}{1.56} = 1.00$$

In this example, the values for the molar ratio at this point are *not* all close to small whole numbers but can be rounded to $3:5:6:1$. Multiplying this ratio by 2 does result in the necessary small whole numbers that are needed for an empirical formula—in this case, $7:12:2$. The empirical formula is therefore $C_7H_{12}O_2$. ●

• • • • • • • • • • •

2 Determining the Molecular Formula

To convert an empirical formula to a molecular formula, we need to know the **molecular weight** of the compound. An analytical laboratory can determine the molecular weight along with the elemental analysis. Freezing-point depression is one way this is accomplished. In many cases, an organic chemist can determine the molecular weight of a compound from its mass spectrum.

In a molecular formula determination, the molecular weight of the empirical formula is simply compared with the experimentally determined molecular weight. If the two weights are the same, the empirical formula represents the molecular formula. If the values are *not* the same, the experimental molecular weight should be a simple multiple of the molecular weight of the empirical formula. This molecular formula is determined by multiplying the empirical formula by this number.

EXAMPLE The theoretical molecular weight of CH_3O is 31.03. The actual molecular weight of the compound in question is found to be 62.11, which is very close to 2×31.03. Therefore, the molecular formula for the compound is $(CH_3O)_2$, or $C_2H_6O_2$. ●

• • • • • • • • • • •

3 Interpreting the Results

In research, an elemental analysis is often used as evidence to substantiate the identity of a proposed structure. In a chemical journal, an analysis might be reported as: "Anal. Calcd for $C_2H_6O_2$: C, 38.70; H, 9.74. Found: C, 38.79; H, 9.67."

Generally, an analysis that results in values within 0.3% of the calculated value for each element is an acceptable piece of evidence for structure proof. An extremely pure sample will usually analyze closer to the calculated values. An analysis that does not fall within 0.3% suggests either an impure sample or possibly a different compound from the one expected.

If a compound contains a large number of carbon and hydrogen atoms (often the case with molecules of biological or medical importance), it is difficult to determine the exact molecular formula from analytical data with any accuracy. For example, *cholesterol*, $C_{27}H_{46}O$, contains 83.87% C and 11.99% H. Cholesterol contains one double bond. When this double bond is hydrogenated, *cholestanol*, $C_{27}H_{48}O$, is obtained. Cholestanol contains 83.43% C and 12.45% H. The percent compositions of these two compounds are fairly close, as you can see—much closer than the percent compositions of comparable smaller molecules would be. It would be difficult to differentiate cholesterol and cholestanol by elemental analysis alone.

APPENDIX III

Toxicology of
Organic Compounds

. . . .

Virtually every chemical has the potential for toxicity. Even dietary necessities can be poisonous when consumed to excess. A few years ago, a Florida woman died from drinking too much water! Most compounds encountered in the organic laboratory are far more toxic than water. Yet, it has been only in the last two or three decades that toxicities of various types of organic compounds have been studied at all. It was not until the 1970s that comprehensive surveys were undertaken. Because of this, and because it is difficult to test toxicities with human subjects, much of the available data is tentative or is applicable directly only to rats, mice, dogs, or monkeys. The allowable, "harmless" levels of organic compounds in the immediate environments of workers and consumers are still being scrutinized by both government agencies and private corporations.

The techniques used in different studies are not always the same. One study may be concerned with the medical symptoms in a group of factory workers. Another study may involve feeding potentially toxic compounds to laboratory animals. Yet another piece of data may come from an isolated case reported in a medical journal.

Other problems that arise in toxicology are the differences between short-term and long-term exposures, individual allergic reactions, variable individual tolerances, and differences in individual habits (for example, does the subject smoke, or is he on any medication?). In time, comprehensive and directly comparable lists of toxicities may become available. Today, we must still extrapolate from one study to another, often with inadequate information.

There are several types of toxicity. A compound may be caustic and irritating, leading to burns and rashes. A compound may be relatively harmless in short-term exposure but may cause cancer if the exposure is repeated or prolonged. A short-term exposure to a compound may be harmless to an adult but cause serious defects in an unborn baby. (For this reason, pregnant women should pay strict attention to safety procedures.) Some compounds are deposited in the fatty tissue of the body instead of being eliminated and so may build up in concentration and lead to chronic toxicity. (This is frequently the case with organohalogen compounds.) The route of administration of a compound also affects its toxicity. *Inhalation* of vapors, aerosols, or dusts usually causes severer symptoms than other types of exposure, such as ingestion or dermal absorption.

Keeping all these variables in mind, let us consider some of the ways that toxicological data are reported. The mode of testing is usually self-evident, or should be; the route of administration and the species tested, as well as the toxicity levels, are a necessary part of any reliable report. For example, the term **LD$_{50}$** (lethal dose, 50% kill) means that half of a statistical sample of test animals died at a particular dose level. The route of administration will be designated as *inhalation*; *oral*; *dermal*; *iv* (intravenous); *im* (intramuscular); *ip* (intraperitoneal, or injected into the abdominal cavity); or *subcutaneous* (injected just under the surface of the skin). In Table III.1, you can see that

TABLE III.1 Toxicities of some common solvents[a, b]

Name	Formula	Reported human toxicity[c]	Maximum allowed occupational level (time-weighted average)[d]
methanol	CH_3OH	inhalation TC_{Lo}, 300 ppm	200 ppm
ethanol	CH_3CH_2OH	oral TD_{Lo}, 50 mg/kg oral LD_{Lo}, 1400 mg/kg	1000 ppm
acetone (propanone)	$(CH_3)_2C{=}O$	inhalation TC_{Lo}, 500 ppm	1000 ppm
diethyl ether	$(CH_3CH_2)_2O$	oral (rat) LD_{50}, 2200 mg/kg	400 ppm
benzene[e]	C_6H_6	inhalation TC_{Lo}, 210 ppm	10 ppm
carbon tetrachloride[e] (tetrachloromethane)	CCl_4	inhalation TC_{Lo}, 20 ppm	10 ppm
chloroform (trichloromethane)	$CHCl_3$	inhalation TC_{Lo}, 10 ppm	50 ppm
methylene chloride (dichloromethane)	CH_2Cl_2	inhalation TC_{Lo}, 500 ppm	500 ppm

[a]H. E. Christensen and T. Luginbyhl, eds. *Toxic Substances List.* Rockville, Md.: U.S. Dept. of Health, Education, and Welfare, 1974.
[b]Ligroin and petroleum ether (both mixtures of alkanes) are not listed, but they are only mildly toxic.
[c]TC_{Lo} is the *lowest published toxic concentration;* TD_{Lo} is the *lowest published toxic dose;* LD_{Lo} is the *lowest published lethal dose;* and LD_{50} is the *lethal dose, 50% kill.*
[d]For comparison, the maximum allowed level of hydrogen cyanide (HCN) is 10 ppm.
[e]Suspected or known carcinogen, can be absorbed through the skin.

the LD_{50} of diethyl ether is 2200 mg/kg when given orally to rats. This means that when diethyl ether was administered through a tube into the stomachs of a statistical sample of rats at the rate of 2200 mg of ether per kilogram of rat body weight, half the rats died. If only one rat from the statistical sample had died, the toxicity would be reported as the *lowest lethal dose,* or LD_{Lo} (oral, rats). If only one rat showed any ill effects at all, then the toxicity would be reported as the *lowest toxic dose,* or TD_{Lo} (oral, rats).

Because inhalation toxicity is of great importance to industrial workers and chemists exposed to fumes and dust, data are often reported as *toxic concentration,* or **TC**. The term TC_{Lo} means the lowest concentration known to cause any toxic effects in one or more members of the group of subjects. The term TD_{50} means, of course, that 50% of the subjects suffered from one or more toxic effects. The *threshold limit value* **TLV** is the maximum amount of vapor in the air that is reported safe for continuous exposure. Concentrations are generally reported as **parts per million (ppm)** or occasionally as **parts per billion (ppb)**. An inhalation concentration of 1.0 ppm is 1.0 μL of sample per liter of air, where 1.0 μL = 0.001 mL or 1.0×10^{-6} L. (As a liquid concentration, the term 1.0 ppm usually means 1.0 mg of solute in 1.0 L of solution.)

Table III.1 lists the reported toxicities of some popular organic solvents. Until a few years ago, *all* these solvents were used regularly in industrial, research, and student laboratories. In the past four to five years, methanol, benzene, chloroform, and carbon tetrachloride have been used less widely because their toxicities have been highly publicized. Today, chemists are more likely to choose ethanol over methanol, methylene chloride over chloroform or carbon tetrachloride, and toluene over benzene.

Table III.2 is a list of some toxic organic compounds. This list is not complete. Many compounds that are *suspected carcinogens* are not included. The list of known or suspected *teratogens* (compounds that can cause fetal damage) is too long to include here; therefore, we have shown only some of the more common ones. The list of *allergens* includes only a few of the compounds that frequently cause allergic reactions; unfortunately, a particular individual may be allergic to a compound that has no effect on most other people.

The fact that many toxic compounds can be *absorbed through the skin* was largely ignored until recent years. For example, a lethal dose of phenol ("carbolic acid"), once used as a surgical antiseptic, can be absorbed dermally. A laboratory worker would be wise to assume that *all* compounds can be dermally absorbed. In many cases, a relatively harmless solvent can carry other, nonabsorbable compounds through the skin. Dimethyl sulfoxide (DMSO), which can actually be tasted by a person after it has been applied to his or her hand, has been used to administer nonabsorbable drugs dermally. This solvent-carrying effect is a good reason not to use a solvent to cleanse your hands of other organic compounds unless absolutely necessary. If a solvent must be used, choose the least objectionable solvent appropriate: ethanol or an alkane solvent like petroleum ether (or, better, mineral oil). Then, scrub immediately with soap and water.

TABLE III.2   Some toxic organic compounds[a]

Known carcinogens[b]	Allergens
acrylonitrile	pyrethrums[d]
4-nitrobiphenyl	diazomethane
α- and β-naphthylamine	p-phenylenediamine
methyl chloromethyl ether	some glues, gums, and resins
benzidine	
4-aminodiphenyl	
1,2-dibromo-3-chloropropane	**Compounds that can be absorbed dermally**
ethyleneimine	methanol
β-propiolactone	1-propanol
2-acetylaminofluorene	1-pentanol
4-dimethylaminoazobenzene	allyl alcohol
N-nitrosodimethylamine	2-chloroethanol
vinyl chloride	dimethyl sulfoxide
benzene	acrylonitrile
4,4-methylenebis(2-	benzene
chloroaniline)	nitrobenzene (and some other
	aromatic nitro compounds
	including 2,4-dinitrophenylhydrazine)
Known or suspected teratogens[c]	bromobenzene
	phenol (and some substituted phenols)
benzene	aniline (and some other aryl amines)
toluene	
xylene	
aniline	
nitrobenzene	
phenol	
vinyl chloride	
formaldehyde	
dimethylformamide	
dimethyl sulfoxide	
N,N-dimethylacetamide	
carbon disulfide	

[a]This list is not intended to be complete. The reader is referred to *Carcinogens*, OSHA 2204, U.S. Dept. of Labor, Jan. 1975; *Safety in Academic Chemistry Laboratories*, 3rd ed., Washington, D.C.: American Chemical Society, 1979; L. Bretherick, ed., *Hazards in the Chemical Laboratory*, 3rd ed., London: The Royal Society of Chemistry, 1981; N. I. Sax, *Cancer Causing Chemicals*, New York: Van Nostrand Reinhold Co., 1981.
[b]A compound that can cause cancer in laboratory animals or humans.
[c]A compound that can cause a physical or functional defect in the embryo or fetus.
[d]A commonly used, naturally occurring insecticide that is isolated from one variety of chrysanthemum.

Suggested Readings

National Research Council. *Prudent Practices for Handling Hazardous Chemicals in Laboratories*. Washington, D.C.: National Academy Press, 1981.

Safety in Academic Chemical Laboratories. Washington, D.C.: American Chemical Society Committee on Chemical Safety.

Registry of Toxic Effects of Substances. Cincinnati: U. S. Department of Health and Human Services, National Institute for Occupational Safety and Health.

Suspected Carcinogens. Cincinnati: U. S. Department of Health and Human Services, National Institute for Occupational Safety and Health.

Aldrich Catalog Handbook of Fine Chemicals. Milwaukee: Aldrich Chemical Co., current edition.

R. Gerlach. "Toxic Chemicals: Understanding TLV's." *J. Chem. Ed.* **1986,** *63(4)*, A100.

H. H. Fawcett, ed. *Hazardous and Toxic Materials*. 2nd ed. New York: Wiley, 1988.

Disposal of Chemical Wastes

. . . .

An important part of laboratory work is being able to safely dispose of the chemical wastes generated by the experiment. Unfortunately, there is no easy or uniformly safe way to do this. The procedures used in any laboratory will depend upon the chemical involved and the geographical location. Some locations are more environmentally sensitive than others. Consequently, the laws and the practices associated with the disposal of chemical wastes are not uniform throughout the world. However, we all agree that *it is the responsibility of the chemist to be aware of and to obey all disposal rules and regulations that apply to his or her laboratory.* If you don't know, ask. Ignorance is not an excuse.

While definite rules that apply to all situations and laboratories cannot be given, there are some general guidelines that do apply. Chemicals cannot be routinely disposed of by dumping them in the sink or wastebasket. Before you start an experiment, be aware of the disposal procedures that apply to the wastes that will be generated by the experiment. This information should be part of the prelaboratory write-up of the experiment.

. . . .

1 Acids and Bases

Mineral acids and bases must be neutralized before disposal. The solution must be diluted and then neutralized slowly to a pH of 7 (or neutral to litmus paper). In most locations, the neutralized solution can be disposed of down the drain.

No generalizations concerning organic acids and bases can be given. Check with your instructor or storeroom personnel concerning disposal of these chemicals.

• • • • • • • • • • •
2 Inorganic Ions

Specific laboratory rules will govern the disposal of heavy metal ions. In general, these ions cannot be disposed of down the drain but, rather, are collected in waste containers and transported to a chemical disposal site.

Dilute solutions of heavy metal ions should be concentrated before the waste is transferred to the waste container. In most laboratories, it is the storeroom's responsibility to collect, package, and ship the concentrate to a safe chemical disposal site.

• • • • • • • • • • •
3 Organic Solvents

Halogenated Hydrocarbons and Hydrocarbon Solvents

In most laboratories, these solvents are collected in specified containers located in the fume hood for recycling or later disposal in a chemical waste disposal site. These solvents must never be disposed of down the drain. As with inorganic ions, it is usually the storeroom's responsibility to collect, package, and ship these waste solvents to a safe chemical disposal site (which, for these solvents, generally involves incineration).

Oxygenated Solvents

Some oxygenated solvents, like ethanol, can be safely disposed of down the drain. Others, like diethyl ether (ether) cannot. Ethanol is readily degraded by bacteria in sewage plants. Diethyl ether, however, is very volatile and floats on water. It collects in sink traps throughout the laboratory and evaporates back into the air.

Do not assume that just because the solvent is oxygenated it can be degraded by sewage bacteria. If you are not positive about how you are to dispose of the chemical, ask your instructor.

Laws, rules, and regulations in the area of waste management are constantly changing as more information becomes available and new disposal procedures are developed. Never assume that the way you did it last time is the way you should do it now. When in doubt, ask.

Suggested Readings

Phifer, R. W., and McTigue, W. R., Jr. *Handbook of Hazardous Waste Management for Small Quantity Generators*. Chelsea, Mich.: Lewis Publishers, 1988.

Gannaway, S. P. "Chemical Handling and Waste Disposal Issues at Liberal Arts Colleges." *J. Chem. Ed.* **1990,** 67(7), A183.

Armour, M. "Chemical Waste Management and Disposal." *J. Chem. Ed.* **1988**, 65(3), A64.

Armour, M., Browne, L. M., and Weir, G. L. "Tested Disposal Methods for Chemical Wastes from Academic Laboratories." *J. Chem. Ed.* **1985**, 62(3), A93.

Aldrich Catalog Handbook of Fine Chemicals. Milwaukee: Aldrich Chemical Co., current edition.

National Research Council. *Prudent Practices for Disposal of Chemicals from Laboratories.* Washington, D.C.: National Academic Press, 1983.

For disposal recommendations of a specific compound, always consult the MSDS (see the Introduction).

Index

···